Food Technology in the Year 2000

Bibliotheca Nutritio et Dieta

No. 47

Series Editor
J.C. Somogyi, Rüschlikon-Zürich

KARGER

Basel · München · Paris · London · New York · New Delhi · Bangkok · Singapore · Tokyo · Sydney

Proceedings of the International Minisymposium on Food Technology in the Year 2000, Helsinki, November 14, 1989

Food Technology in the Year 2000

Volume Editors
S. Lindroth, Turku
S.S.I. Ryynänen, Helsinki

18 figures and 17 tables, 1990

KARGER

Basel · München · Paris · London · New York · New Delhi · Bangkok · Singapore · Tokyo · Sydney

Bibliotheca Nutritio et Dieta

Library of Congress Cataloging-in-Publication Data
International Minisymposium on Food Technology in the Year 2000 (1989: Helsinki, Finland)
Food technology in the year 2000: proceedings of the International Minisymposium on Food Technology in the Year 2000. Helsinki. November 14, 1989 / volume editors, S. Lindroth, S.S.I. Ryynänen.
(Bibliotheca nutritio et dieta: no. 47)
Includes index.
1. Food industry and trade — Congresses. I. Lindroth, Seppo, 1950– . II. Ryynänen, S.S.I.
(Suvi Sisko Irmeli), 1960– . III. Title. IV. Series.
[DNLM: 1. Food Technology — congresses.]
ISBN 3-8055-5243-2

Bibliographic Indices
This publication is listed in bibliographic services, including Current Contents® and Index Medicus.

Drug Dosage
The authors and the publisher have exerted every effort to ensure that drug selection and dosage set forth in this text are in accord with current recommendations and practice at the time of publication. However, in view of ongoing research, changes in government regulations, and the constant flow of information relating to drug therapy and drug reactions, the reader is urged to check the package insert for each drug for any change in indications and dosage and for added warnings and precautions. This is particularly important when the recommended agent is a new and/or infrequently employed drug.

ISBN 3-8055-5243-2

Contents

Preface

The Food Research Program of the Finnish Ministry of Agriculture and Forestry is a temporary governmental research body with the mission of promoting the diversification and internationalization of food production in Finland. It was set up in 1984 and its term will end on April 30, 1990.

The research objectives of the Program have been to raise the competitiveness of Finnish food production, based on domestic raw materials and know-how, both in Finland and internationally. The research in the Technology Projects of the Program has focused on new methods of creating and maintaining excellent quality in foods (e.g. freeze-drying, modified atmosphere packaging and storage, high frequency heating and decontamination of foods and food processing environments by gamma or ultraviolet irradiation) as well as on biotechnology in food production. In addition, promoting international cooperation in science and technology has been a key activity.

Although Finland is a small country in population, with only five million people, her food research, food, and related technology and equipment manufacturing in certain specific fields are both prominent and recognized worldwide. Examples of such areas are biotechnology, dairy technology, and production and distribution of ready-to-eat chilled foods.

The effective utilization of existing technology and the application of newly developed methods and processes play key roles in the optimal production of food to meet consumers' changing demands and preferences. As a concluding seminar to the Technology Projects of the Food Research Program, the International Minisymposium on Food Technology in the Year 2000 was organized. The objective of the symposium was to assess future needs, priorities and trends in the area of food technology research through presentations by, and active discussions among experts in the field. For practical reasons, the participation was limited to 40, on an invitational basis. Since we feel the presentations merit further distribution to other specialists and experts in the field, we have submitted the proceedings of the Symposium for publication.

Seppo Lindroth
Chairman of the Symposium
Planning and Organizing Committee

Welcoming Words

Juhani Kuusi

Technology Development Center, Helsinki, Finland

Recent developments in technology and especially the extension of these developments to embrace all sectors of society, have greatly affected every industrial sector – the food industry included – and every industrialized country – including Finland.

Before going deeper into the food industry and food technology, let me shortly describe the Finnish response to the galloping development of technology.

Finland, like many other industrial countries, is currently going through a period of significant industrial and social restructuring.

To date, adaptation to the change in the world economy has proceeded rather smoothly. Economic growth has been steady and exports have been diversified.

While forest industry products accounted for 75% of the exports in 1960, the figure in recent years has been around 40%. The metal industry together with the engineering and electrical industries now account for the same percentage as the forest industry.

In 1988, 64% of all exports went to the EC and EFTA countries of Western Europe.

There has been broad agreement in Finland that success in coping with this period of change will depend greatly on the science and technology policy pursued. In 10 years the research and development percentage of the Finnish GNP has increased from 1% to nearly 2%; the goal for the year 2000 is 2.7%.

It is significant that the private sector contributes more than 60% of the total amount. The role of industry in research and development is therefore considerable. One of the most important features in the Finnish science and technology policy in the current decade has been the rapid expansion of international cooperation.

Active participation in multilateral Western Europe projects such as EUREKA, the EC research programs and the European Space Agency have been prominent in this respect.

Finland has been involved in EUREKA since its foundation in 1985. The agreement offering Finnish companies and organizations an opportunity to participate in the EC research programs was signed in April, 1986.

Finland has been an associate member of the European Space Agency since 1987; the Finnish Government recently voiced its support for the initiation of talks aimed at full membership by the beginning of 1995.

What are the goals of the Finnish technology policies today?

For the present stage of development of technology and world trade, Finland's technology and industrial policy should, in particular, aim at: (1) the maintenance of the competitiveness of the basic industries (forest industry, metal and engineering industry), and (2) the creation of new industry in the developing technology sectors.

These aims are not contradictory. The maintenance of the competitiveness of the basic industries also requires rapid efficient utilization of new technologies – comprehensive technological innovation. An important condition for this is that developing technology sectors should have their own research, development and industrial activities. Without these, the correct utilization of the latest technological developments in the sphere of basic industries will be difficult or impossible.

Thus, we in Finland shall have a strong horizontal interaction between basic industries and new industries based on developing technology sectors such as information technology, biotechnology, and materials technology.

Everything I have said before concerning basic industries is applicable to the food industry.

The food industry is, by volume and by the number of employees, one of the largest industrial areas in all countries. In Finland the food industry comes directly after the wood and metal industries. The food industry consists of both large – in many cases international – companies and small regional companies. It is practiced all over the country. Its products are sold directly to consumers, not to the industries of other areas. So everybody has more or less direct contact daily with the food industry and the technology behind it.

Like all of the basic industries, the food industry and the technology it uses need continuously renewing injections from developing new technology sectors to maintain competitiveness and to be able to meet the ever-increasing demands of consumers for healthier products with low energy value, less or better fat, more natural or less processed products, etc.

Biotechnology has traditionally been practiced in the food 'industry': cheese, beer and wine production is based on the use of microorganisms. Biotechnology has also been utilized in bread making (baking). However, new applications of biotechnology or biotechnical methods will find use in food technology: new, more efficient or better microorganisms and metabolites and/or tailored raw materials will be developed by using modern biotechnical methods such as gene technology, protein engineering, and plant biotechnology. A part of the applications of these new methods in industrial production may be delayed because of possible negative images, especially those linked with genetic engineering of animals or with the use of foreign hormones in animal feeding.

The food industry has to be aware of ethical discussions concerning the utilization of these new technologies. In practice, people working in these areas must be able to participate in these debates by presenting facts, not emotional feelings.

Though biotechnology is unquestionably important to the food industry, the importance of process technology shall not be forgotten. The development of this important area shall be carried out in cooperation between food technologists and equipment manufacturers.

A low degree of automation is typical of the major part of the food industry. One reason for this is the lack of suitable measuring systems. The knowledge of what really is happening to the raw material during processing is in many cases minimal. Basic and applied research is needed here.

The need for automation will grow in the future. The food industry has to cut the costs in all possible ways. There will be keen competition for people who are willing to work within the industry. The only way to overcome this problem is to add automation where it is possible. Well-defined automation processes will also help the industry to satisfy new demands for healthier and less processed food.

The latest results of other developing technology sectors should similarly be effectively applied within the food industry. To attain this, increasing national and international R&D activities are needed.

At the national level, additional problems are caused by a very great number of small- and medium-sized companies which have very limited resources for carrying out research work or even for following the course of technology development. Here, active and communicative public research centers and laboratories may benefit a lot.

It is pleasing to note that international R&D cooperation is increasing in this field.

European cooperation in food technology has been running under several COST projects with 19 countries participating in the research work. There are at least three EC research programs related to food technology, namely FLAIR, ECLAIR and BRIDGE. Within EUREKA, biotechnology in particular has an important role.

Nordic cooperation has been running in some independent projects but will be strengthened under a new biotechnology umbrella program, food biotechnology. This program will include cooperation between different Nordic research groups, industry and national financing bodies.

We see that the food industry is facing new demands from many different points of view. This also means new challenges for those working within food technology. These challenges become emphasized in the intimate and direct contacts with end consumers. We have a lot of evidence of this fact in Finland today.

The Food Research Program of the Ministry of Agriculture and Forestry has been one of the tools with which we have created a basis for meeting these challenges. The program is now in its final phase. This seminar organized by the Food Research Program is a very important part of the termination and summing-up activities.

I hope this symposium will offer ideas about what is expected to be essential in food technology in the future and guidelines for further activities to meet increasing future challenges.

Invited Participants

Prof. R.K. Abrahamsen
Agricultural University of Norway
Norway

Ms. Maarit Ahola
Ministry of Agriculture and Forestry
Finland

Mr. Clas Bergring
Imatran Voima Oy
Finland

Dr. Martin Checa
Conservera Campofrio, S.A.
Spain

Dr. Sampsa Haarasilta
Vaasamills Ltd.
Finland

Mr. Seppo Heiskanen
Finnish Food Industries' Federation
Finland

Prof. Jorma Hirn
National Veterinary Institute
Finland

Prof. Lea Hyvönen
University of Helsinki
Finland

Mr. Tapani Kivelä
Hackman-MKT Oy
Finland

Mr. John B. Klis
Institute of Food Technologists
USA

Mr. Johan Korkman
Farmers' Union
Finland

Mr. Heikki Kupari
ETEK Oy
Finland

Dr. Rauno Lämsä
Ministry of Finance
Finland

Mr. Raul Lönnström
Valio Engineering Ltd.
Finland

Mr. K.B. Madsen
Danish Meat Research Institute
Denmark

Mr. Seppo Mäkinen
SITRA
Finland

Prof. Yrjö Mälkki
Technical Research Centre of Finland
Finland

Dr. K.M.K. Muttzall
Unilever Research Laboratory
The Netherlands

Dr. Matti Näsi
Agricultural Research Center
Finland

Ms. Paula Nybergh
Technology Development Center
Finland

Dr. Thomas Ohlsson
SIK – The Swedish Food Institute
Sweden

Dr. Takeyoshi Ohsawa
Asahi Foods Corporation
Japan

Mr. Markus Onnela
Ministry of Social Affairs and Health
Finland

Prof. Kaisa Poutanen
Technical Research Centre of Finland
Finland

Prof. Dr. Ernst H. Reimerdes
Meggle Milchindustrie GmbH & Co KG
FRG

Mr. Harald Relander
Oy Lahden Polttimo Ab
Finland

Ms. Suvi Ryynänen
Ministry of Agriculture and Forestry
Finland

Ms. Eva Tornberg
Swedish Meat Research Institute
Sweden

Dr. Gerhard Wittl
Cerveceria Polar C.A.
Venezuela

Lindroth S, Ryynänen SSI (eds): Food Technology in the Year 2000.
Bibl Nutr Dieta. Basel, Karger, 1990, No 47, pp 1–5

Future Developments and Research Priorities in Food Technology

Esko H. Koskinen

Food Research Program, Ministry of Agriculture and Forestry,
Helsinki, Finland

Food technology has its beginning in prehistoric times. Technologies for baking, brewing, butter and cheese manufacture, wine making, etc. were known thousands of years ago. They were developed, however, by accidental discoveries or random trials, not by systematic experiments and at any rate not by scientific methods. We might attribute scientific food technology to Pasteur and his cooperation with the distillers of Lille about 130 years ago. Since then, scientists have had an increasing role in the development of processing, manufacture and preservation techniques for food.

For a long time, food technology was considered a cluster of independent technologies in brewing, dairying, etc. Only about 50 years ago was it realized that commodities were not the essentials but that certain principles were valid regardless of the origin or purpose of the materials processed. Food technologists became organized according to related methods and techniques used in food processing. Now we are speaking of food technology, which comprises engineering, biotechnology, sanitation, instrumentation, material development, processing technology, packaging technology and storage technologies. In these fields food technology is closely associated with other sciences.

What will the future requirements for food technology and the development trends be? How will they affect the research priorities in food technology? Let's take a look at the food industry, which should be seen as a part of the food chain from agriculture and fisheries to consumers. In this chain, basic production, especially agriculture, was the leading line of business until World War II, but it was rapidly displaced in developed countries by the food industry. Nowadays various service functions like distribution, retailing, catering and restaurant services tend to dominate the competition for the consumer's food expenditures. For a long while, the success factor has not been the possession of raw materials, but production technology. Now it seems that marketing know-how has become the success factor of today. Simultaneously, the degree of freedom in business has advanced from a closed economy to an open international competition.

The food system exhibits a remarkable shift in the nature of innovations towards the consideration of consumer's new life styles. Technology development seeks to provide benefits from needs newly revealed by modern concepts in the life sciences. The rate of innovation concerning new plants or animals which could be used as food decreased long ago. There also seems to be a decline in the rate of process innovations as we have already mastered the economy of traditional food processes reasonably well. New process innovations still emerge, especially in the field of biotechnology. The applications of existing processing technology to local foods creating new varieties for international use or for special purposes are rapidly increasing.

This means that there is now a growing sophistication about technological innovations in the food industry. In almost saturated markets, where the consumption of food energy need not be increased, fulfilling the consumer's demands for tastier and safer food with more variety, convenience and psychological value will be increasingly important. The profitability of companies seems to vary more with the effectiveness of their marketing and market research than with their technology. Technological innovation is only one route to growth, and it may not be the most effective. It is no longer possible to easily maintain dominance in the market only by technological innovations when competitors wait until a new process or product has been developed or marketed and then move in either through purchase or through a closely competitive development. Scientific research is only one route to technological innovation. Will the ability to quickly apply results of research carried out elsewhere be more important than maintaining a large staff for a company's own research? There are many companies whose research departments are devoted mainly to technological intelligence.

How far can we go? How long will there be an ample supply of research results from independent research institutions? It is already apparent that governments have to accept new responsibilities for scientific and technological research to ensure support for the food system, because the needs to improve the quality of food products require a lot of scientific and technological innovations. Food technologists are expected to solve various problems connected to new varieties of food with improved processing capability, new foods and new sources of foods, new processing methods, and new packaging methods and materials, as well as with continuous processes. Innovations are demanded from them for developing convenience products, flexible processes or improved distribution systems. Inquiries are made about improved quality of food products, increased stability, prolonged shelf life, etc.

Although the food system contributes significantly to the gross national product in developed countries, it is structurally very complex and fragmented. There are few large industries which spend substantial amounts on research, and the vast majority of food processors operate on a small scale and small profit margin. These processors depend on national governments and research institutions when scientific and technological support is needed. Demand for such support is greatest in small market areas where the processor has to improve the quality of his products rather than increase the quantity produced in order to ensure his foothold. Small companies could have good prospects in the special products market, but more know-how is needed for developing such products than for producing standard products.

The interests of the society in general in supporting the development of food technology are aimed at improving food safety and quality for the consumer, at increasing productivity, at diminishing environmental pollution and at ensuring proper utilization of agricultural and fishery products. Locally, the objectives of governments are also to promote food industry competitiveness and to regulate international trading. In recent years there has appeared an increasing consumer interest and requirement for healthy, nutritious food and demand for convenience, greater variety and hedonistic satisfaction.

The environment in which food technology research is done comprises universities, research institutions and enterprises. Research is regulated by funding from national budgets, private science- and technology-promoting foundations, and enterprises. Priorities are set by scientists themselves, governmental agencies and financiers.

Universities have three roles. They educate scientists and technologists, they conduct research and they provide industry with services. Universities are laying the foundation for future innovation by training the future generation to carry out basic and applied research. Graduates usually take positions in industry and government agencies, and those awarded a doctor's degree are also employed by universities and research institutes. The new technologies pose enduring problems: How to train enough teachers and how to keep them at the universities when they can at least double their salaries by moving into the commercial sector?

In addition to basic research training, universities should provide opportunities for the continuing education of food technologists to keep them abreast of recent developments in the field. This is possible only when basic research programs are carried out by university scientists. Applied research can also be carried out by universities. It has been suggested that applications become more and more important to universities. Up-to-date education needs modern equipment and instruments, which could also be effectively used for research aiming at industrial applications. Many universities have advisory programs for assisting industry in making use of new research information to improve their products or processes or to reduce costs.

Research institutes should concentrate on applied research with a practical mission. It is, however, necessary to carry out some basic research, especially on methodologies, in order to maintain a high level of know-how among the employees. Private research institutes would emphasize research connected with product or process development and would service industry. A government institute may practice serving enterprises, but, in addition to general applied research in areas of nationally strategic importance, they are mainly involved in regulatory programs related to safety and hygiene of food processing and to programs administering standards for processes or grades for foods. The work done by private institutes differs from that done in government institutes mainly in regard to greater emphasis on creating new products or processes, whereas the research in government institutes tends to be more thorough and systematic, with the main emphasis on publishing the results.

The research of food technology in enterprises can be on various levels. The research idea entered the industrial scene as a service to production, as the company's first effort to understand and regularize its production process. Gradually, in advanced companies of ample resources, research has gained the status of an independent entity, and research policy is considered one of the central development strategies of the company. Research in technology has also tended to yield unexpected results. Companies in the food business find themselves in possession of new devices for extruding plastics, for example. In most cases, however, the degree of sophistication has been low. Usually the quality questions, economy or introduction of a new processing method for facilitation of manufacture of a new product line have stirred up the demand for research. It has mainly been problem-solving research, and when a problem was identified the laboratory and pilot research may have been carried out in a private research institute to reach a practical solution.

National science- and technology-promoting organizations shoulder much of the burden of defining priorities in research. They can allocate resources to nationally important subjects of research, and their decisions have directed activities in universities and research institutes towards a new field. Communication between enterprises and these organizations has too often been poor, and an improved course of action is needed for ensuring common aims and approaches of government and industry.

This leads me to stress the need for information and the necessity for communication for the effective advancement of food technology research. We need more interaction in defining research priorities, and we need it even more in search for adequate resources. Know-how also includes knowing how to use knowledge and resources available elsewhere. There is a need for national efforts, but there are also reasons for transnational initiatives for cooperation programs in food technology. In my opinion, priority should be given to closer international collaboration. I would like to emphasize a more effective use of resources achieved by new kinds of partnerships between universities and research institutes as well as with industry.

Esko H. Koskinen, ScD, Director, Provivo Oy, Tekniikantie 17,
SF–02150 Espoo (Finland)

Lindroth S, Ryynänen SSI (eds): Food Technology in the Year 2000.
Bibl Nutr Dieta. Basel, Karger, 1990, No 47, pp 6–11

New Opportunities in Food Technology Created by Biotechnology

Tor-Magnus Enari

Technical Research Centre of Finland (VTT), Espoo, Finland

Introduction

Biotechnical methods have been applied for thousands of years in the production of fermented foods, various dairy products and alcoholic beverages.

The development of molecular biology has recently opened new possibilities for food technology as well [1]. New biotechnical methods which can be or have been used in food technology are: genetic engineering; protein engineering; enzyme technology; immobilization of cells and enzymes; plant cell culture techniques.

These new techniques can be applied in the development of agricultural products used as feedstock in the food industry; industrial production processes; analytical methods and safety and quality assessment; new food products.

Feedstock for the Food Industry

The food industry uses mainly agricultural products as raw material. Recent developments in biotechnology have opened up new possibilities for both plant breeding and breeding of farm animals.

Plant Breeding

Novel genetic techniques are to some extent already used in plant breeding but their impact will increase in the near future as a result of development and application of the new techniques. It is presently not possible to transfer genes to monocotyledonous plants. This means that large groups of agricultural plants, such as all cereals, still cannot be transformed using recombinant DNA techniques [2]. Within a few years, however, the problems will almost certainly be overcome and the new techniques will help in the solving of many problems.

Most research groups active in this field are interested in agricultural problems [2] such as: disease resistance; herbicide resistance; resistance to weather conditions; improved yield; decreased dependence on fertilizers.

Much less attention has been directed towards properties of the crop. This is, however, a question which calls for the special attention of the food industry [3, 4]. Such properties are: improved nutritional value; lysine-rich cereals; altered fatty acid composition.

And: improved technical properties; improved baking quality of wheat; improved filtrability of brewer's wort; improved shelf stability of tomatoes; improved texture; improved flavor.

Some of these requirements are controlled by only one or a few genes. Such alternatives will soon be possible with the aid of recombinant DNA technology. Other aims are more complicated and their solution must await basic research into plant metabolism and physiology.

Animal Breeding

Transgenic animals have already been made using recombinant DNA technology. However, the present possibilities are still limited and the ethical questions are under debate. The questions presented for this technique are mainly confined to: nutritional properties; growth and reproduction; disease resistance.

Food Ingredients

Many food ingredients can be produced by fermentation processes with bacteria, yeasts or fungi but also by plant cell cultures [5, 6]. The production organisms can be improved using genetic techniques. Food ingredients produced by biotechnical processes will in many cases replace chemical products which have been banned for health reasons.

Food ingredients which are or will be produced by biotechnical methods are: colors; flavors; sweeteners; preservatives; antioxidants; vitamins; thickeners; emulsifiers; texturizers.

Production Processes

In food production many biotechnical processes are traditionally used. In these processes it is only natural that the new possibilities created by the development of molecular biology will also be used. Furthermore, the new techniques will be used in many processes as a substitute for or complement to traditional chemical or physical processes.

Starters

Microbial pure cultures have been used for only about a hundred years in industrial processes. They have slowly replaced the natural and often mixed populations used in the production of alcoholic beverages; baker's yeast; dairy products; fermented meat products; fermented vegetable products.

Genetic engineering has been used to improve the properties of yeast strains [7] and lactic acid bacteria. Research on the improvement of brewer's yeast has reached a stage where the yeast is technically ready for industrial application [8]. Presently, however, no brewery has used these yeasts in their production. The largest scale of applications has been 50 liters.

The reason for the hesitation in using the new techniques is fear of consumer reactions. The technical and economic advantages are nevertheless so great that the techniques will certainly be used as soon as they are generally accepted.

Fermentations

Fermentations are traditionally batch processes but a continuous process would in many cases lead to improved production economy and a more consistent product. For these reasons research work on continuous processes has been intense during the last 25 years. The immobilization of microorganisms and enzymes has opened up new possibilities for the development of continuous processes [9, 10].

Industrial applications of immobilized cells in food fermentations are scarce. This is, of course, due to the complexity of the product. When simple single compound chemicals, such as ethanol, are produced, the process is easy to optimize and control. When the product is a complex mixture in which the degree of fermentation, flavor formation and texture should be controlled, a traditional batch process has the advantage of automatically yielding the traditional product. Here the difficulties in employing continuous processes are based on insufficient knowledge of the nature of the production process.

Enzyme Technology

Enzymatic reactions participate in many food processes [11, 12]. The endogenous enzymes of the raw material may have an essential role in the production process but they may also have a detrimental effect on the product. A traditional example of enzyme technology is the malting of barley. Here the enzymes needed in beer production are produced by germinating the barley. In many materials the enzymes cause spoilage, color changes or development of off-flavors.

The addition of enzymes has also been used for a long time in food processes. Plant enzymes include the proteolytic enzymes papain, ficin and bromelin. The chymosin used in cheese production is of animal origin. Many microbial enzymes have been used for years.

Enzymes which have been used in industrial processes are mainly extracellular hydrolytic enzymes. Commercial enzyme preparations are usually mixtures of several enzymes although one activity normally dominates. Side activities may influence the process although the commercial preparation is sold under a specific name.

Modern biotechnology has widened the field of application of enzymes. Recombinant DNA technology has made it possible to produce enzymes free of unwanted side activities. The technique has also enhanced yields in enzyme production and made it possible to produce enzymes which were previously too expensive for industrial use. Immobilization techniques provide a means for reusing enzymes or designing continuous processes. This has a significant influence on process economy. Finally, protein engineering will soon enable us to design enzymes with new or altered properties [13, 14].

Analyses, Safety and Quality Assessment

Several new methods have recently been developed for analytical purposes. These methods include methods for assessment of toxic substances and infectious microorganisms. Biotechnical methods for these purposes include enzymatic methods, monoclonal antibodies, DNA probes and new immunological tecniques.

Conclusions

The rapid development of new biotechnical methods will to a large extent also influence the food industry and its processes. The food industry has applied biotechnology in its processes for thousands of years but the new methods which have emerged from molecular biology were used first in the pharmaceutical industry. This is due to many factors. Firstly, the pharmaceutical industry is much more active in research and development work whereas the food industry is rather conservative. Secondly, pharmaceutical products are much more expensive and can thus better carry the increased processing costs of a more sophisticated technology. Thirdly, the consumer is rather conservative in his choice of food products but when a cure is needed for an illness he looks at all possibilities.

The feedstock of the food industry is, nevertheless, biological material. It is thus only natural to use biological methods for its processing. The methods of the new biotechnology are also biological methods and consequently it is logical that they should be used in food technology.

References

1 Pelsy G: The impact of biotechnology on the agricultural and food industry. Biofutur June 1989;80:10–18.

2 Van Montagu M: Genetic engineering of plants for agriculture. EBC Symp Plant Biotechnology, Helsinki, 1989, monogr XV (in press).

3 Enari T-M: Plant biotechnology for the malting and brewing industries. EBC Symp Plant Biotechnology, Helsinki, 1989, monogr XV (in press).

4 Teeri T: Improvement of malting barley. EBC Symp Plant Biotechnology, Helsinki, 1989, monogr XV (in press).

5 Bell ERJ, White EB: The potential of biotechnology for the production of flavours and colours for the food industry. Int Ind Biotechnol 1989;9:20–26.

6 Drawert F, Berger RG: The use of microorganisms and plant cell cultures for the production and biosynthesis of aroma substances. EBC Symp Biotechnology, Nutfield, 1983, pp 122–143.

7 Knowles JKC, Tubb RS: Recombinant DNA: Gene transfer and expression techniques with industrial yeast strains. EBC Symp Brewer's Yeast, Helsinki, 1986, monogr XII, pp 169–185.

8 Enari T-M: Genetic engineering of industrial yeasts – a brewer's approach. Yeast 1988;4(special issue): 28.

9 Linko M: Immobilized enzymes and cells. Proc 20th Congr Eur Brewery Convention, Helsinki, 1985, pp 39–50.

10 Setzermann U: Immobilized biocatalysts in the food industry. Lebensmittelindustrie 1988;35:103–106.

11 Christensen FM: Enzyme technology versus engineering technology in the food industry. Biotechnol Appl Biochem 1989;11:249–265.

12 Linko M: Enzymes in the forefront of food and feed industries. Food Biotechnol 1989;3:1–10.

13 Creamer LK, Jimines-Flores R, Richardson T: Genetic modification of food proteins. Trends Biotechnol 1989;6:163–169.
14 Leatherbarrow J, Fersht AR: Protein engineering. Protein Eng 1986;1:7–16.
15 Knorr D: Biotechnical processes in food production. Food Biotechnol 1987;1:225–248.

Prof. Tor-Magnus Enari, Research Director, Division for Process Technology,
Technical Research Centre of Finland,
Vuorimiehentie 5, SF–02150 Espoo (Finland)

Lindroth S, Ryynänen SSI (eds): Food Technology in the Year 2000.
Bibl Nutr Dieta. Basel, Karger, 1990, No 47, pp 12–18

Integration of Hygiene into Food Technology

Esko Nurmi, Jorma Hirn

National Veterinary Institute, Helsinki, Finland

The primary task of food hygiene is the prevention of food-borne diseases in humans, mainly disorders caused by infections, toxins or chemicals. Further hygienic measures are applied in order to attain a high quality of the completed food. For this purpose, hygienic measures have to be applied through the whole production chain, starting with the animals or other basic food source and continuing until the final product reaches the consumer. Only from the high-quality raw materials can high-quality products be produced.

Quality of the foods, especially those of animal origin, is a very extensive and complex question. With the quality we may mean functional, nutritional, sensory or hygienic quality of the food. Hazards to human health, which are often discussed in connection with, for example, meat, milk and poultry products are: (1) infections and intoxications; (2) residues and contaminants; (3) food additives.

When we speak about residues and contaminants in the completed food, we have to agree that little has to be gained by different methods of food technology. The best way to control these hazards is to avoid raw materials containing such substances. On the contrary, we are able to influence and reduce the hazard caused by microbes by combining the methods of hygiene and technology.

A dramatic increase in food-borne bacterial diseases has been observed. Only half a century ago we knew of three bacteria transmitting diseases through various foods, i.e. *Clostridium botulinum, Staphylococcus aureus* and *Salmonella.* Until today we have learned about some 20 pathogenic species or groups of organisms, with more emerging all the time. During the last 10–20 years *Campylobacter* spp., verocytotoxin-producing *Escherichia coli, Listeria monocytogenes, Salmonella enteritidis* and *Yersinia enteroco-*

litica have become prominent. The significance of *Aeromonas hydrophila* and *Plesiomonas shigelloides* in the etiology of food-borne infections has yet to be clarified.

There is an agreement among authorities and research workers that food-borne infections have increased remarkably during the last years, e.g. in USA and Great Britain human *Campylobacter* infections are today more predominant than *Salmonella* infections. Most countries report an increase of *Listeria* and *Yersinia* cases. Furthermore, *S. enteritidis* has become the most common serotype as a cause of human salmonellosis in countries like the USA and Finland. The organism seems to be closely associated with eggs, e.g. in England and USA and may become established in the ovaries of hens. In the meeting organized by the WHO in Genova during March 20–23, 1989 [1], one important statement arrived at was: 'The present status of *Salmonella* contaminations of poultry products is not acceptable and every effort must be made to provide and market *Salmonella* free products.' The participants were unanimous in stressing the importance of introducing effective veterinary hygienic measures in agricultural production. The difficult but, nonetheless, achievable goal is farms free of *Salmonella* infection.

There may be many reasons for the increase in food-borne infections, among others the liberalized trade in feed and foodstuffs, international travel, animal mass production and modern technology. In addition, food hygiene is either neglected or in need of improvement in many countries. There are also certain changes in food habits which may lead to food-borne infections or intoxications. As examples we can mention nonpasteurized milk and various milk or meat products but also so-called natural foods.

It is very easy to say 'Keep pathogens out of food.' As already mentioned, during the last 5–10 years, in most countries we have had a dramatic increase of food-borne infections. It is therefore a reason to question what the situation will be in the year 2000.

Food Processing and Supply in the Year 2000

The food habits of the consumer will presumably change a great deal before the end of the century. The consumer becomes more health conscious and greater demands will be made upon the quality, freshness and safety. Food production and refining become more variable. We will get more semimanufactured foods prepared at special shops. Ready-made foods and also imitation products and foods will find even higher markets. The consumers' attitude towards the use of chemicals in animal mass production and food processing is becoming reserved and more negative. New technology and the alteration of habits and expectations of the consumer make increasing demands on food hygiene. Since it is often a matter of a pathogen shared by animals and man, the protective measures should start with animal production. Food-borne enteropathogens are more and more often identified in food-producing animals and we may ask why? In the same way we may ask if our food technology and processing have changed during the last decades. The biggest changes may be constituted by the enormous increase in the use of cold chains and the constant demand to reduce food additives.

Control of Food-Borne Infections and Intoxications

Most of the food-borne infections have their origin in the animal production. Which are the methods then to control these infections and intoxications? The means of control can be grouped in the following way: (1) production of pathogen-free animals; (2) prevention of the environmental contamination of raw materials and foods; (3) developed processing technology combined with the elimination methods; (4) information and education of the food handlers and consumers; (5) research.

Research Related to the Control of Food-Borne Infections and Intoxications at the National Veterinary Institute in Finland

It is well known that poultry is the most important source of *Salmonella* and *Campylobacter* infections in humans. Therefore, we have at the National Veterinary Institute first concentrated our research on the prevention of these pathogens in the broiler production. The traditional preventive methods are either cost prohibitive or practically impossible to take. Therefore, methods based on new ideas have been sought. One of the most promising is that of competitive exclusion, which is based on a Finnish discovery published in *Nature* in 1973 [2]. We demonstrated that oral introduction of intestinal bacterial flora from adult birds into newly hatched chickens increased the chicks resistance to *Salmonella infantis* infections. This approach became the subject of numerous investigations. It could be demonstrated in the Finnish and other studies that this protective bacterial flora was cultivable. The method is called competitive exclusion (CE) or the Nurmi concept and has been used in Finland since the year 1976. It is also in use in Sweden and England.

The culture initially used in Finland was an anaerobically cultured broth culture from adult birds. Several research groups have tried to develop mixtures of pure cultures as we also have done at the National Veterinary Institute in Helsinki, but today there is no product based on pure cultures available. The only commercial product on the market is produced in Finland by the name Broilact®, which is based on a manipulated, selected anaerobically cultivated pathogen-free broth culture. The usual treatment is given with the first drinking water, but, for example, spray application in the hatcheries has also been shown to be effective.

The actual and reported field experiments have been relatively limited in number. The greatest experience concerning the effects of this method on *Salmonella* is to be found in Finland and Sweden. During the years 1982–1987 nearly 5 million broilers have been treated in Sweden. The epidemiological evaluation strongly supports the fact that the culture has a *Salmonella*-preventing effect also under field conditions [3].

In Finland the method has been used for over 10 years and today over 70% of the broilers are treated by Broilact®. In summer 1987, for example, one slaughter house chain was found to have about 50% of their 55 broiler flocks infected. By improved hygiene and the Broilact® treatment the share of infected flocks before slaughtering dropped to less than 10%. Another broiler slaughter house chain has, during a 3-year period, followed the effect of Broilact® comparing the *Salmonella* infection percentage between treated and nontreated flocks. During the period from 1.1.1986 to 31.12.1988 they treated 400 broiler flocks by the CE product Broilact® and left 192 without treatment. Of the treated flocks only 26 were *Salmonella* positive and of the nontreated flocks 40 were *Salmonella* positive at the time of slaughter. According to statistic comparison the treatment had significantly reduced the number of *Salmonella*-infected flocks. Furthermore, the research at the National Veterinary Institute has shown that even though the birds treated with the CE method are carriers of *Salmonella* in the intestines, the content of bacteria is very low. When studying deep-frozen broiler carcasses marketed in Finland in 1989 we found 5.1% positive for *S. infantis.* The number of *Salmonella* per carcass was very low varying from 2 to 7 salmonellae per kg. The decrease of the number of salmonellae in the chicken intestines has in itself great importance for food hygiene and public health.

The success in protecting chicks against salmonellae by competitive exclusion has encouraged us and some other working groups to apply the same method on other enteropathogens, especially *Campylobacter.* Today it appears feasible that comparable preparations against *Campylobacter* can be developed. Also pathogenic serotypes of *E. coli* are under research.

Food Additives and Food-Borne Pathogens

The use of preservatives is a very old custom when we consider salt and saltpeter, for example. A lot of research has also taken place regarding the preserving effect of these additives. During the last 20 years various consumer organizations and legislative bodies have suggested a reduction of additives in various foods in general and table salt, nitrate and nitrite especially. In our opinion the other side of the matter has often been forgotten or ignored, i.e. the effect of these additives on the growth of various food-borne pathogens. In the literature there are remarkably few studies explaining the combined effects of various additives in foods on the growth of pathogens. Such kinds of studies have been carried out at the National Veterinary Institute in the 1970s and 1980s. These studies have mainly been concentrated on meat products like dry sausages and cooked sausages, but also other kinds of foods like bakery products and fish products have been studied. Our earlier studies revealed that the growth of *Salmonella* [4], *S. aureus* [5] and *Y. enterocolitica* [6] may be prevented in salami sausages by combining the starter culture, nitrite or nitrate and table salt. In vacuum-packed cured trout we could demonstrate the combined effect of nitrate and table salt in reducing the toxin formation of *Clostridium botulinum* [7]. One of the most interesting facts was when we could demonstrate that using glucono-delta-lactone (GDL) as a pH-value-reducing additive in bakery filling composed of butter, eggs and sugar, the growth of *Salmonella* was inhibited. According to our results, the prevention of the growth of pathogens largely depends on rapid decline of the pH value caused by starter cultures and GDL together with additives like nitrite and nitrate and also on reduced water activity (a_w) caused by table salt [8].

Only recently the bacterium called *L. monocytogenes* has reached a significant position in the field of food hygiene. It has proved to survive or even reproduce in salami manufactured by modern technology. Therefore, we wanted to study the effects of the technology and additives used some 30 years ago, on the *L. monocytogenes* bacteria. In the 1950s the manufacturing of salami in Finland included 1,000 ppm of saltpeter and 3.5% of table salt. In the 1960s the amount of nitrite allowed was, in Finland, some 200 ppm and the addition of salt was 3.5%. At the moment the permitted addition of sodium nitrite is 120 ppm and the amount of salt usually added is 3.0%. Our experiments showed that *L. monocytogenes* survived in all additive concentrations except the one used for salami manufacturing in the 1950s. The content of this bacterium in the latter salami went down to one thousandth during the 3-week ripening period [9].

In our studies of cooked sausages we concentrated on the combined effects of various additives. We have, for example, found out that a liver sausage with a long preservation period shows a strong growth of *Bacillus cereus, C. botulinum* and *C. perfringens* in the summertime [10]. Our research revealed that with an appropriate combination of various additives, which all are permitted, the growth of *B. cereus* and *C. perfringens* could be prevented. This effect of sodium nitrate on the prevention of bacterial growth was greatly increased when the use of nitrite was combined with the use of GDL decreasing the pH value (with 0.2–0.3 pH units), and with the use of sodium erythorbate [11].

These studies have served to show that the policy aiming at the decrease in the use of salt and other additives affecting the preservation, makes the growth of pathogens possible in several meat products. On the other hand, the appropriate combinations of various additives even in permitted amounts may be used for controlling the concentrations of pathogens in various foods.

Recommendations for the Food Processing Industry

The food industry is faced with more concerns about the microbiological safety of products.

Newly recognized food-borne pathogens occur and food-borne infections are increasing.

We have to aim to reduce the occurrence of food-borne pathogens like *Salmonella, Campylobacter* and *Yersinia* in our food-producing animals. The goal must be, for example, a broiler production free of *Salmonella* or at least a much lower percentage of *Salmonella* and *Campylobacter* contaminated birds on the market. Also, the amount of these pathogens is important.

In the food industry major changes have to be made in the formulating, packaging and storage of foods.

Food manufacturers should consult much more with food hygienists and microbiologists to review the changes and innovations they intend to introduce for their products.

It is extremely important to study the combined effect of different food additives, varied concentrations and combinations on pathogenic bacteria before decisions of the reduction of food additives are made. Inoculation studies with potential problem micro-organisms may be necessary to evaluate the safety of such products.

Consumption of raw foods like eggs from *S. enteritidis*-infected areas is not to be recommended. Also, the use of raw eggs as in mousse, creams or ice cream, that do not receive a heat treatment or addition of additives to reduce the growth of *Salmonella* should be avoided.

Finally, we must agree that food-borne infections cannot be eradicated, but we have to make efforts together in order to get our food as safe as possible.

References

1 World Health Organization: Report of WHO consultation on epidemiological emergency in poultry and egg. WHO/CDS/VPH/89.82., 1989.

2 Nurmi E, Rantala M: New aspects of *Salmonella* infection in broiler production. Nature 1973;241:210–211.

3 Wierup M, Wold-Troell M, Nurmi E, et al: Epidemiological evaluation of the *Salmonella* controlling effect of a nationwide use of a 'competitive exclusion'-culture in poultry. Poultry Sci 1988;67:1026–1033.

4 Sirviö M, Nurmi E, Puolanne E, et al: Der Einfluss von Starterkulturen und verschiedenen Zusatzstoffen auf das Wachstum von *Salmonella senftenberg* in der Rohwurst. Fleischwirtschaft 1977;57:1007–1012.

5 Niskanen A, Nurmi E: Effect of starter culture on staphylococcal enterotoxin and thermonuclease production in dry sausage. Appl Environ Microbiol 1976;31:11–20.

6 Raevuori M, Nurmi E, Hill P: The effect of nitrate, nitrite and a starter culture on the growth of *Yersinia enterocolitica* in dry sausage. Proc 24th Eur Meet Meat Research Workers, Kulmbach, 1978.

7 Ala-Huikku K, Nurmi E, Pajulahti H, et al: Effect of nitrite, storage temperature and time on *Clostridium botulinum* type A toxin formation in liver sausage. Eur J Appl Microbiol 1977;4:134–149.

8 Nurmi E, Koiranen L, Raevuori M: The use of glucono-delta-lactone for the prevention of salmonellosis and staphylococcal food-poisoning due to bakery fillings. 7th Int Symp WAVFM, Garmisch Partenkirchen, 1977.

9 Junttila J, Hirn J, Hill P, et al: Effect of sodium nitrite and potassium nitrate in the survival of *Listeria monocytogenes* during the ripening of dry sausage. J Food Prot 1989;52:158–161.

10 Ala-Huikku K, Nurmi E, Pajulahti H, et al: The occurrence of *Clostridium botulinum* type E in Finnish trout farms and the prevention of toxin formation in fresh-salted vacuum-packed trout fillets. Nord Vet Med 1977;29:386–391.

11 Asplund K, Nurmi E, Hill P, et al: The inhibition of the growth of *Bacillus cereus* in liver sausage. Int J Food Microbiol 1988;7:349–352.

Prof. Esko Nurmi, National Veterinary Institute,
PO Box 368, SF-00101 Helsinki (Finland)

Lindroth S, Ryynänen SSI (eds): Food Technology in the Year 2000.
Bibl Nutr Dieta. Basel, Karger, 1990, No 47, pp 19–37

Food Technology Research Plan for the Year 2000 in the Japanese Agriculture, Forestry and Fisheries Research Council Secretariat to Assist the Food Industry to Maintain and Improve Its Competitive Edge

Takashi Sasaki[1]

National Food Research Institute, Ministry of Agriculture, Forestry and Fisheries, Tsukuba, Japan

Basic Policies of MAFF for Promoting Research on Food Technology Towards the 21st Century

Basic Policies

In 1986, the Agricultural Policy Council concluded that policies relating to the food industry should be enriched in order to update Japanese agricultural policies towards the 21st century. More particularly, the Council suggested examining the general course of the Japanese food industry by taking into consideration changes in the structure of demands including the diversification of needs in our eating habits (fig. 1), and the recent internationalized situations such as international location of companies and an increase in the importation of processed products.

Next, it was indicated that national research institutes should conduct studies in the following important fields.

(1) Public fields (standards for the safety of food additives, measures to counter public pollution and environmental problems, establishment of authorized analytical methods, etc.).

(2) Fields urgently needed to satisfy social needs (techniques for saving energy and resources, etc.).

[1] I would like to acknowledge Director-General and Dr. T. Nisio, International Research Division and Biotechnology Division of Agriculture, Forestry and Fisheries Research Council for their valuable advice.

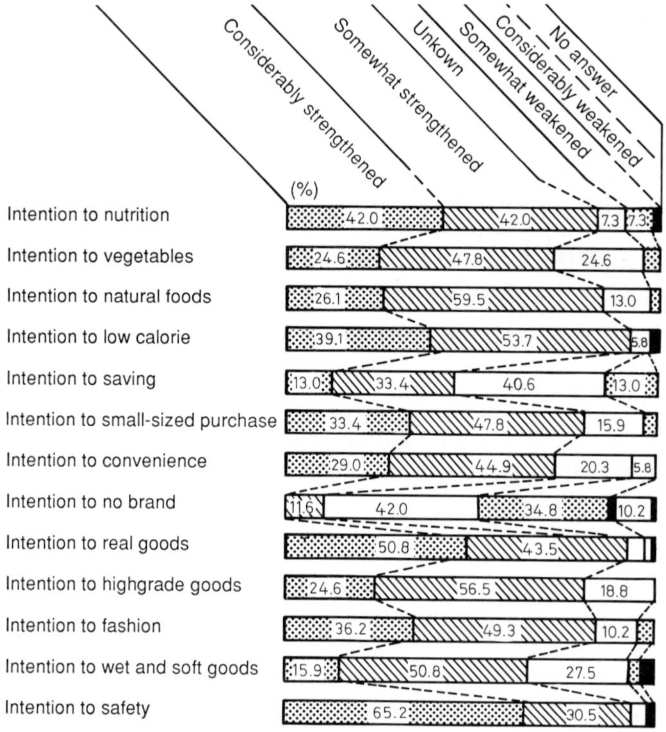

Fig. 1. Estimation of changes in consumers' needs in Japan (Japan Committee for Economic Research). It is estimated that needs relating to healthy life such as intentions to safety (95.7%), real good (94.3%), low calorie (92.8%), natural foods (85.6%) and nutrition (84.0%) will be elevated.

(3) Fields to be evaluated objectively (evaluation of safety, nutritional evaluation, etc.).

(4) Excessively risky fields for private companies (utilization of unused resources such as food materials, development of new food materials, etc.).

(5) Fields to be organized and promoted by the government (up-to-date techniques such as genetic engineering).

(6) Basic and widely applicable fields (packaging technology, marketing technology, preservation technology, research on food manufacturing machinery, etc.).

(7) Fields scarcely studied in private institutes (improvement in traditional fermented foods and processed marine products, processing and utilization of Japanese crops such as rice, etc.).

1987 Estimation of Technical Development Towards the 21st Century in Agriculture, Forestry and Fisheries Research

The Council Secretariat carried out a questionnaire on the estimation of the development of food technology towards the 21st century to research into the extent of completion and the popularization of techniques which will be finished by 1995. Table 1 shows the results where ⊙, ○ and ● show the opinions of 70% or more, 50–70% and 50% or less of the subjects taking part in the questionnaire, respectively.

These results suggest the following points.

(1) Each unit operation procedure is thoroughly evaluated not only in the extent of completion but also of popularization. However, some show an extremely high extent of completion but a considerably low extent of popularization (table 1, points 1), 6) and 21)).

(2) Item relating to sensors and the evaluation of qualities and safety show a relatively high extent of completion. However, the extent of popularization of these items is low.

(3) In the field of biotechnology, the application of enzyme engineering shows a high extent rate both of completion and of popularization. However, the extent of completion of the application of biotechnology, such as cell fusion, recombination of DNA and bioreactors, is generally low. Further, the extent of popularization of these items is, needless to say, also low.

These facts indicate that research in the field of biotechnology, which potentially involves unlimited possibilities, should be prior to others in order to further develop food technology towards the 21st century. And also figure 2 shows the problems to be solved by the Government toward the 21st century.

Recent Trends in Biotechnological Research

Private Companies

According to a survey conducted by the Division of Biotechnology of the Agriculture, Forestry and Fisheries Research Council Secretariat in 1984, 179 companies, which corresponded to 1/3 of the total, conducted R&D by using biotechnology. These companies include 78 which manufacture drugs, 65 which manufacture foods and drinks, and 58 related to agriculture, forestry and fisheries.

From the viewpoint of R&D costs, the scale for biotechnology in Japanese private companies has certainly continued to grow at a ratio of 20% per year.

Table 1. Results of a questionnaire on the completion and popularization of techniques towards the 21st century

Content of techniques	Completion	Popularization
1 Application of membrane		
1) Efficient recovery of valuable components from food industry waste water	⊙	●
2) Separation of biogas, etc. by using a gas-separation membrane	●	
3) Isolation of isomers from natural compounds	●	
2 Molding and processing		
4) Remolding of scarcely or not utilized protein materials	⊙	⊙
5) Detection of contaminants contaminating during processing	⊙	○
3 Drying		
6) Drying hardly dried foods	⊙	●
7) Efficient drying technique by using novel energy sources	○	●
4 Pasteurization		
8) Pasteurization of solid or liquid mixtures by UHT pasteurization	⊙	○
9) Various long-life food products obtained by aseptic packaging	⊙	⊙
5 Packaging		
10) Selective permeability and selective insulation for definitive gases	⊙	○
11) Convenient detection of pin-holes in package containers	⊙	⊙
6 Freeze-grinding or freeze-concentration		
12) Efficient grinding of highly fibrous food materials	⊙	○
13) Efficient grinding of highly oily food materials	⊙	○
7 Utilization of super critical gas		
14) Acceleration of reaction in nonaqueous system	●	
15) Development of quantitative determination analysis (chromatography)	○	○
16) Efficient separation and extraction of definite useful components	⊙	○

Table 1. (cont.)

Content of techniques	Completion	Popularization
8 Refrigeration, freezing and preservation		
17) Preservation of vegetables and fruits under suppressing physiological activities	○	●
18) Freezing and preservation of hardly freezable foods	●	
19) Development of various cold chain systems	⊙	○
9 Transportation system		
20) Transportation in pressure-reduced containers	○	●
21) Aseptic transportation system for liquid foods	⊙	●
22) Bulk transportation in large-size containers	⊙	○
10 Sensors and automation		
23) Unattached food processing by utilizing mechatronics	⊙	●
24) Development of qualitative components (components and taste) and manufacturing process	○	●
11 Evaluation of qualities		
25) Convenient evaluation of freshness of foods	○	●
26) General evaluation of components and tastes by nondestructive analysis	○	●
27) Development of nutritional evaluation system for eating mode and food composition	○	●
12 Maintenance of safety		
28) Convenient determination of contamination with harmful micro-organisms	○	●
29) Convenient identification of heavy metals, agricultural chemicals, food additives and harmful materials	○	●
30) Convenient detection of mycotoxin contamination and removal of the toxicity	●	
31) Chemical-free insect pest control	○	●

Table 1. (cont.)

Content of techniques	Completion	Popularization
13 Application of enzyme engineering		
32) Continuous enzymatic saccharification of raw starch	⊙	○
33) Continuous enzymatic saccharification of cellulose	○	●
34) Production of new functional foods (sweetener, water-retention agent, etc.) from starch	⊙	○
35) Mass production of novel cyclodextrin	⊙	○
36) Synthesis of sucrose		
37) Development and utilization of synthetic enzymes		
15 Bioreactor		
38) Application of bioreactor to the production of sake, wine, etc.	○	●
39) Application of bioreactor to the production of miso, soy, etc.	○	●
40) Application of bioreactor to the production of acetic acid, lactic acid, citric acid, etc.	○	●
16 Biotechnology including recombination of DNA, cell fusion, tissue culture, etc.		
41) Modification of improvement of enzymatic function by recombination of DNA	●	
42) Improvement of microbial function by the recombination of DNA	○	●
43) Production of reserve protein or reserve starch originating from plant by highly functional microorganism		
44) Production of new useful fungi through cell fusion	●	
45) Production of salt- and osmotic pressure-resistant yeast through cell fusion	●	
46) Stable production of natural colorant and spice by tissue culture	○	●

Data obtained from 'Questionnaire on the estimation of technical development' by Agriculture, Forestry and Fisheries Research Council Secretariat.
⊙ = 70% or more; ○ = 50–70%; ● = 50% or less.

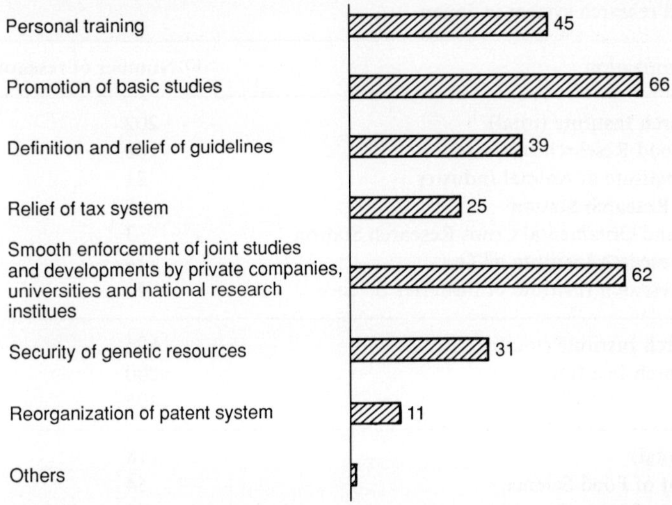

Fig. 2. Problems to be solved by the government (data obtained from Nikkei Bi-Tech). Results (expressed in %) of a questionnaire on problems by which the government is confronted in the promotion of biotechnology. A remarkably large number of subjects expected much from national research institutes and universities on items of 'Promotion of basic studies' and 'Smooth enforcement of joint studies and developments by private companies, universities and national research institutes', compared with other ones. Therefore, it may safely be said that the results indicate the public demand for the government's policy towards the 21st century.

Most food manufacturers conduct biotechnological research relating to fermentation and utilization of enzymes to produce foods, enzymes or physiologically active substances.

Almost 70% of the food manufacturers intend to reinforce their research systems. 132 private companies, which corresponded to about 70% of the total, promoted biotechnological research in cooperation with other institutes such as universities, other companies and private research institutes. Subjects of the research cooperation include almost the same numbers of recombinations of DNA, cell fusions, tissue cultures and bioreactors. Cooperative research by a private organization with another private one most frequently relates to the development of bioreactors, while that by a private company with a university or a national research institute most frequently is concerned with the recombination of DNA.

Forty-six companies, which corresponds with slightly more than 20% of the total given answers (193), conduct research under cooperation with

Table 2. Food research system in Japan

Research organization	Number of researchers
MAFF Research Institute (total)	202
National Food Research Institute	108
National Institute of Animal Industry	21
Fruit Tree Research Station	8
Vegetable and Ornamental Crops Research Station	3
National Research Institute of Tea	18
National Research Institute of Fisheries Science	44
Public Research Institute (total)	605
Food Research Institute	200
Others	405
University (total)	218
Department of Food Science	54
Department of Agricultural Chemistry	95
Department of Agricultural Engineering	16
Affiliated Institute	14
Public University	8
Private University	31
Private Institute	7,633

foreign companies. The recombination of DNA is most commonly observed as the subject of this research.

Universites

About 220 staff members (table 2), including professors and assistant professors, are engaged in food research at universities in Japan, and most of them belong to the faculty of agriculture and conduct basic research relating to the whole food science. This research may be classified into fields connected with food science such as food chemistry, nutritional science and food hygiene, those related to biotechnology such as applied microbiology, those related to the utilization of agricultural products such as production of agricultural products, and those related to food technology. Recently, workers in chemical engineering have attempted an approach to food science.

About 25% of the total research subjects (280) relate to so-called biotechnology, namely, utilization of microbial and enzymatic functions and recombination of DNA.

Although a lot of research seems basic, some research is observed to aim at the production of materials, for example, 'a system for efficiently producing cheese by using an enzyme', and 'the production of useful oligosaccharides by a lytic enzyme'. Further, it is interesting that some research aims at clarifying the nutritional and physiological functions or the normal properties of food. Examples of these research topics include 'analysis on physiological functions of nutrients by using a cell diversification model system under incubation', 'a novel antimutagenic polypeptide produced by a microorganism', and 'an animal culture cell mutant requiring unsaturated fatty acids'. Furthermore, 'systematic analysis on food functions' is a large-scale project which aims at individually clarifying various functions of food such as the physiological functions of peptides and oligosaccharides as well as creating food products with new functions. More than 80 workers in total have participated in this project within 3 years at a cost of about US $4,300,000.

Public Research Institutes

About 210 public research institutes conduct studies relating to biotechnology in Japan (table 2). In the field of food, about 40% of the total prefectures conduct research in order to improve production processes and develop new food materials, and it is estimated that this ratio will increase to 79% in the future. Examples of the subjects of this research include the improvement of sake yeast by cell fusion and the development of bioreactors for the production of wine and vinegar from fruits produced in their own area.

The total cost for biotechnological research in public research institutes amounts to about US $7,800,000 (1984), which corresponds to about US $160,000 per prefecture on average. About 10% of this cost is used for research on food. The number of researchers all over Japan amounts to about 700, which corresponds to an average of about 16 per prefecture. Similar to private companies, public research institutes vigorously conduct research cooperation with other organizations. Namely, almost 30% of public research institutes conducted cooperative, trusted or entrusted research. The most frequently observed partners in these research topics are public research institutes (more than 50%), followed by universities, national research institutes, and private companies.

Every institute plans to conduct a personal training system to cope with the ever-increasing demand for R&D in biotechnology. That is, public research institutes are planning to send their staff members to national research institutes (about 50%) or universities (about 30%).

National Research Institutes (table 2; fig. 3, 4)

In 1984, the Division of Biotechnology was founded in the Agriculture, Forestry and Fisheries Research Council Secretariat in order to vigorously promote biotechnological research. Laboratories conducting biotechnological research at present under the control of MAFF amount to about 18% of the total.

In the field of food, the National Food Research Institute takes the initiative in biotechnological research together with the National Institute of Animal Industry and the National Research Institute of Fisheries. At the National Food Research Institute, about 40% of the total laboratories conduct biotechnological research. The main subjects of research include: (1) detection, identification, classification and conservation of microorganisms relating to food; (2) improvement in the productivities and properties of useful enzymes through the application of novel techniques such as genetic recombination and cell fusion; (3) novel techniques for the production of useful materials by microorganisms or enzymes, and (4) energy-saving efficient production of bioreactors, etc. (fig. 5).

The Plan for Biotechnology R&D of MAFF towards the 21st Century

Objectives

Biotechnology in the fields of agriculture, forestry, fisheries and the food industry is strongly expected to lead these industries to future prosperity through great increases in productivity. Therefore, basic and advanced R&D should be emphasized to develop or improve industrial techniques as well as to contribute to the efficiency of the research process (fig. 2).

In particular, promotion policies for the extension of biotechnology should be gradually encouraged as the technology develops and as more and more research goals are attained. Furthermore, other policies like financing and taxation are also indispensable for the development in biotechnology.

All policies mentioned above are planned and will be carried out systematically.

Fig. 3. Research organization in the MAFF.

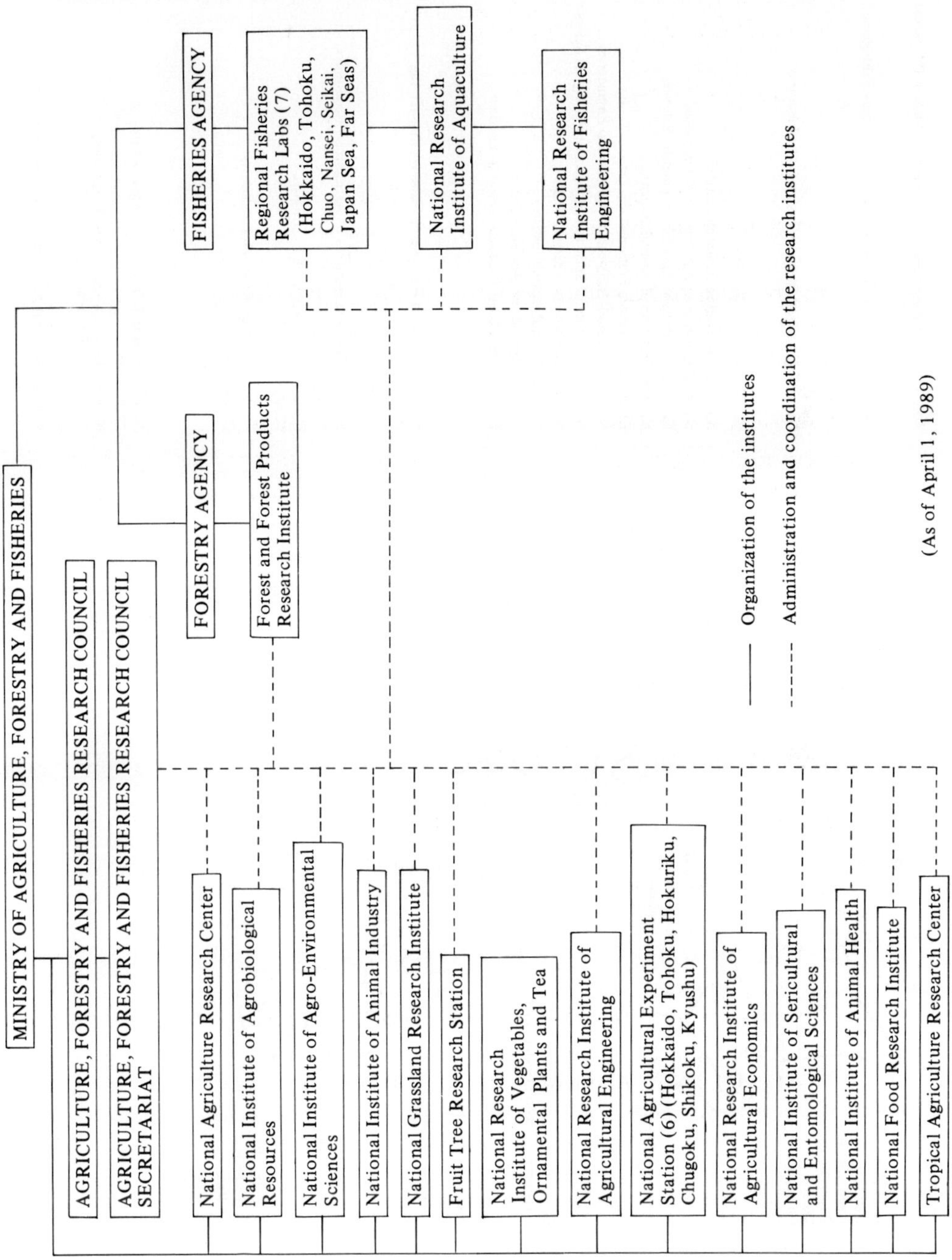
MINISTRY OF AGRICULTURE, FORESTRY AND FISHERIES
AGRICULTURE, FORESTRY AND FISHERIES RESEARCH COUNCIL
AGRICULTURE, FORESTRY AND FISHERIES RESEARCH COUNCIL SECRETARIAT
FORESTRY AGENCY
Forest and Forest Products Research Institute
FISHERIES AGENCY
Regional Fisheries Research Labs (7) (Hokkaido, Tohoku, Chuo, Nansei, Seikai, Japan Sea, Far Seas)
National Research Institute of Aquaculture
National Research Institute of Fisheries Engineering
National Agriculture Research Center
National Institute of Agrobiological Resources
National Institute of Agro-Environmental Sciences
National Institute of Animal Industry
National Grassland Research Institute
Fruit Tree Research Station
National Research Institute of Vegetables, Ornamental Plants and Tea
National Research Institute of Agricultural Engineering
National Agricultural Experiment Station (6) (Hokkaido, Tohoku, Hokuriku, Chugoku, Shikoku, Kyushu)
National Research Institute of Agricultural Economics
National Institute of Sericultural and Entomological Sciences
National Institute of Animal Health
National Food Research Institute
Tropical Agriculture Research Center
Organization of the institutes
Administration and coordination of the research institutes
(As of April 1, 1989)

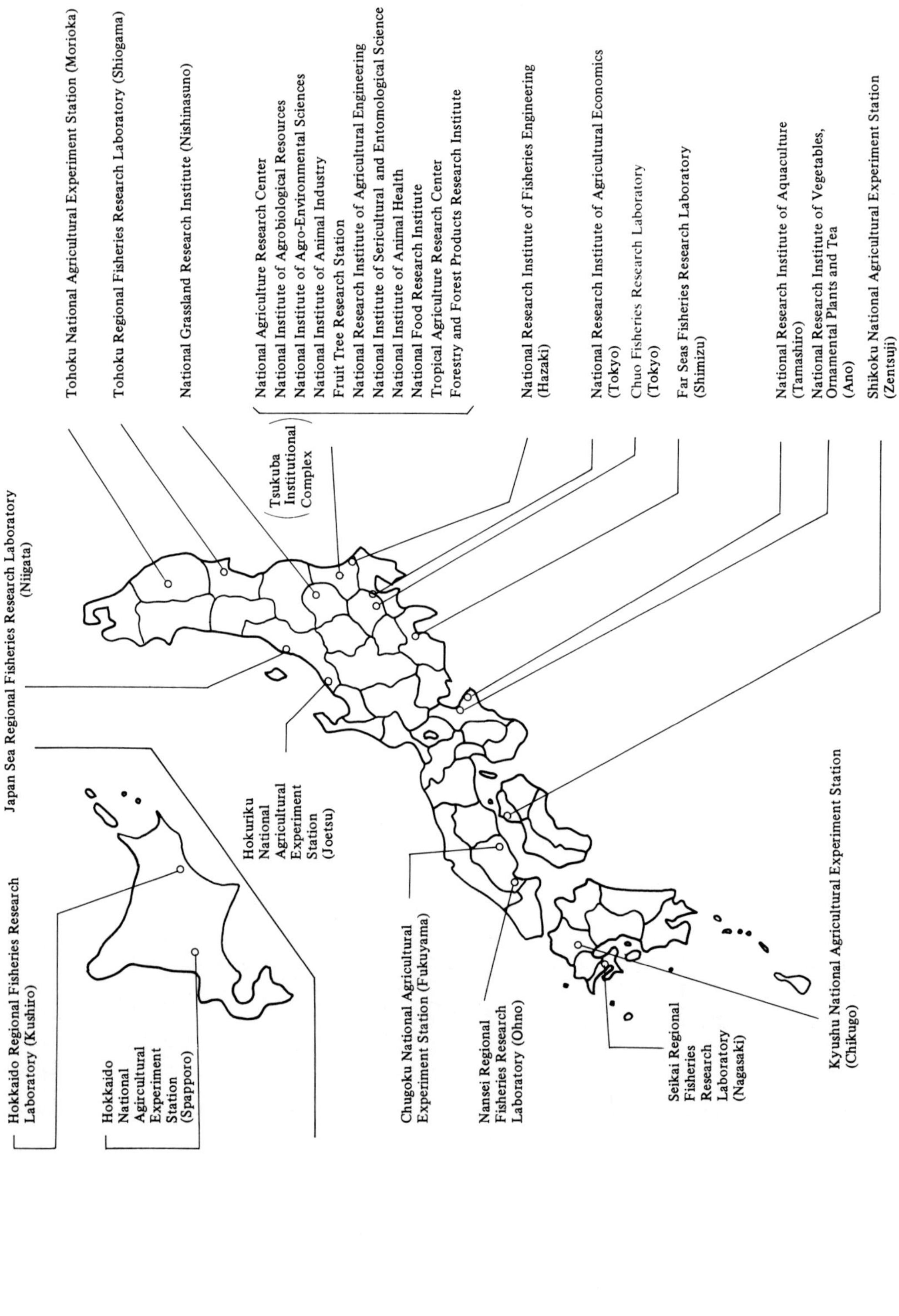
Hokkaido Regional Fisheries Research Laboratory (Kushiro)
Hokkaido National Agricultural Experiment Station (Spapporo)
Japan Sea Regional Fisheries Research Laboratory (Niigata)
Hokuriku National Agricultural Experiment Station (Joetsu)
Chugoku National Agricultural Experiment Station (Fukuyama)
Nansei Regional Fisheries Research Laboratory (Ohno)
Seikai Regional Fisheries Research Laboratory (Nagasaki)
Kyushu National Agricultural Experiment Station (Chikugo)
Tohoku National Agricultural Experiment Station (Morioka)
Tohoku Regional Fisheries Research Laboratory (Shiogama)
National Grassland Research Institute (Nishinasuno)
Tsukuba Institutional Complex
National Agriculture Research Center
National Institute of Agrobiological Resources
National Institute of Agro-Environmental Sciences
National Institute of Animal Industry
Fruit Tree Research Station
National Research Institute of Agricultural Engineering
National Institute of Sericultural and Entomological Science
National Institute of Animal Health
National Food Research Institute
Tropical Agriculture Research Center
Forestry and Forest Products Research Institute
National Research Institute of Fisheries Engineering (Hazaki)
National Research Institute of Agricultural Economics (Tokyo)
Chuo Fisheries Research Laboratory (Tokyo)
Far Seas Fisheries Research Laboratory (Shimizu)
National Research Institute of Aquaculture (Tamashiro)
National Research Institute of Vegetables, Ornamental Plants and Tea (Ano)
Shikoku National Agricultural Experiment Station (Zentsuji)
(As of April, 1, 1989)

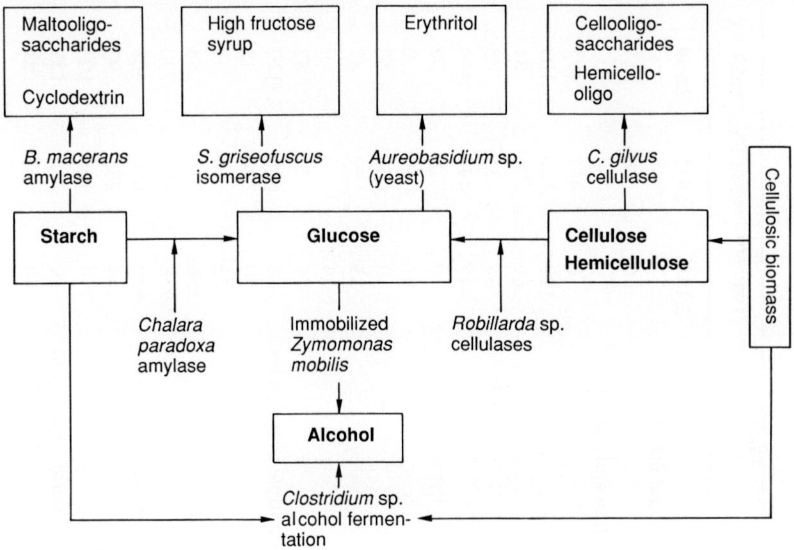

Fig. 5. Topics of carbohydrate biotechnology in the National Food Research Institute (MAFF).

Biotechnology R&D will be encouraged according to the following programs:

(1) Basic and advanced R&D leading towards the 21st century.

(2) R&D for innovative techniques applied to the food industry and other agro-related industries.

(3) Strengthening of the foundation for technological transfer to practical use.

(4) International cooperation in biotechnological research.

(5) Bio-related large-scale programs.

(6) Technological extension at the practical stage.

The total budget for biotechnology R&D has been increasing steadily since the Biotechnology Division was initially created in the MAFF and biotechnology was adopted as the principal strategy of the MAFF's R&D policies in 1984. This budget, 7,425 million yen, increased by 13% in 1989 compared with 6,588 million yen in the previous year.

Fig. 4. Locations of research institutes.

	Title of program	1987 FY	1988 FY	1989 FY
I	Basic and advanced R&D leading towards the 21st century	2,128	2,425	2,944
a	Comprehensive program for the control of ecosystems in agriculture, forestry and fisheries (Bio-cosmos program)	0	0	417
b	Analysis and control of bio-information systems for technological innovation in agriculture, forestry and fisheries (Bio-media program)	0	467	477
c	Basic research for future prospects of biotechnology (entrusted to universities)	100	100	103
d	Molecular biological studies of plants, especially on rice	69	79	81
e	Structure of useful genes in organisms utilized for agriculture[1]	36	34	33
f	Comprehensive research program of biotechnological plant breeding	449	448	458
g	Transformation of plant cells through the introduction of small organelles into the cells[1]	0	9	39
h	Cell fusion for the improvement of microorganisms and plants[1]	41	40	0
i	'New-hybrid' breeding system[1]	0	53	52
j	'Bio-nursery' system	33	39	47
k	Mass propagation of plant seedlings[1]	54	0	0
l	Indentification and utilization of useful animal genes	71	73	82
m	New systems for feeder beef cattle production by in vitro fertilization and multiple pregnancies	0	99	110
n	Breeding techniques including gynogenesis of fish and shellfish	20	20	18
o	Safety assessment of rDNA organisms in open environments	22	38	39
p	Promotion of biotechnology R&D in regional areas (subsidized to prefectures)	256	256	258
q	Management of the 'Gene Bank Project in the MAFF', etc.	978	681	730
II	R&D for innovative techniques applied to the food industry and other agro-related industries	659	650	638
a	Bioreactor system for the food industry[1]	244	235	92
b	Improvement of enzyme functions for the food industry[1]	110	106	103
c	Food production by large-scale high-density fermentation under ultra-high pressure conditions[1]	0	0	139
d	New-type organic fertilizers produced by biological methods[1]	40	48	51

Table 3. (cont.)

	Title of program	1987 FY	1988 FY	1989 FY
e	Basic techniques for the production of agricultural chemicals[1]	58	57	0
f	Efficient production of agricultural chemicals by biosynthesis[1]	0	0	52
g	Immunological reactions for the diagnosis of infectious diseases of livestock[1]	39	38	0
h	Genetically manipulated vaccines against animal protozoa diseases[1]	0	0	37
i	Production and utilization of biomass resources from agriculture, forestry and fisheries	63	72	73
j	Interrelations among roots, soil and soil microorganisms in the rhizosphere of crops and their effective utilization	95	95	90
III	Strengthening of the foundation for technological transfer to practical use	13	19	35
IV	International cooperation in biotechnological researches	5	9	20
V	Bio-related large-scale programs	1,899[2]	2,174[2]	2,353[2]
VI	Technological extension at the practical stage	1,210	1,777	2,328
Total		5,857	6,588	7,425

[1] Subsidized to the private sector.
[2] Some part of (V) is doubly counted.

As shown in table 3, this program is intended for plant, animal, forestry, fisheries and food. The food project program is as follows.

Main Research Programs

Basic and Advanced R&D Leading towards the 21st Century

(1) Analysis and control of bio-information systems for technological innovation in agriculture, forestry and fisheries (Bio-media program). Elucidation of various biological functions like those of physiological activators or environmental stimulation to realize the maximum potential of organisms.

(2) Basic research for future prospects of biotechnology (entrusted to universities). Basic, innovative and multidisciplinary research by universities for developing future possibilities in biotechnology. The research subject in 1989 is enzyme design.

(3) Cell fusion for the improvement of microorganisms and plants. Development of the applicable techniques of cell fusion, including protoplast preparation, actual fusion and hybridoma culture, for the improvement of microorganisms and plants.

(4) Safety assessment of recombinant DNA organisms in an open environment. Development of methods for pre-assessing the safety of recombinant DNA organisms in ecosystems.

(5) Promotion of biotechnology R&D in regional areas (subsidized to prefectures). Development of techniques for the improvement and utilization of local biological resources in collaboration among national and prefectural research institutes. The research subject in 1989 is application of microorganisms to food processing.

(6) Management of the 'Gene Bank Project in the MAFF'. Project for the collection and conservation of genetic resources, application management of genetic information, including the exploration of genetic resources in foreign countries, the systematization of data-base management, etc.

R&D for Innovative Techniques Applied to the Food Industry and Other Agro-Related Industries

(1) Bioreactor system for the food industry. Development of bioreactors using immobilized microorganisms or enzymes and of its application to food-processing systems.

(2) Improvement of enzyme functions for the food industry. Structural elucidation and improvement of natural and artificial enzymes for industrial use.

(3) Food production by large-scale high-density fermentation under ultra-high pressure conditions. Improvement of enzyme functions for efficient food production by applying ultra-high pressure conditions to fermentation.

(4) Production and utilization of biomass resources from agriculture, forestry and fisheries (Biomass conversion program). Techniques for the conversion and utilization of biological resources through the application of microorganisms and enzymes.

Strengthtening of the Foundation for Technological Transfer to Practical Use

(1) Methodology for the extension of newly developed biotechnology in regional areas. Research and case studies for the extension of newly developed biotechnology to practical use by the application of regional biological resources.

(2) Personnel training programs for researchers. (a) Intensive training program on rDNA technology for researchers in national institues. (b) Training program for private and prefectural researchers by conducting research at national research institutes for 2–9 months.

International Cooperation in Biotechnological Research

(1) Basic and advanced research by international cooperation. Basic and advanced 3-year research programs by international cooperation and international workshops for information exchange in biotechnology.

(2) Research cooperation with postdoctoral researchers. Research cooperation mainly on rDNA technology with postdoctoral researchers from developing countries.

Bio-Related Large-Scale Programs

Large-scale R&D programs through the application of results in biotechnology, including the construction of common facilities for international and domestic research cooperation. Themes are as follows: super rice program; biomass conversion program.

Other Relevant Policies

Bio-Oriented Technology Research Advancement Institution (BRAIN)

BRAIN is the semi-governmental institution invested by the government and private sector, established in October 1989. The role of BRAIN is to promote bio-oriented research carried out by the private sector. Its main functions are as follows:

(1) Investment in newly established bio-venture companies (up to 70% of total capital).

(2) Financing to existing companies.

(3) Coordinating cooperative research between national institutes and the private sector.

(4) Assisting the access of the private sector to the 'Gene Bank System'.

Collection and Conservation of Genetic Resources

We currently have genetic resources from plants (130,000), animals (600), microorganisms (8,000), forest trees (21,000) and fishes and sea vegetables (230) in the 'MAFF Gene Bank' (1987 FY). A project for the collection, classification, identification, characteristic evaluation, conservation and management of genetic resources is in progress. A distribution service was started in January, 1986 for the resources from plants, and in September 1987 for those from microorganisms.

Construction of new facilities began in 1985, and was completed in March, 1988 to respond to the large increases and diversification of the collection of genetic resources. These facilities are located in the National Institute of Agrobiological Resources, and are expected to function as a key station for the MAFF Gene Bank Project and also for the field of advanced research in biotechnology.

Research Cooperation

(1) Science and technology (S&T) agreements. Japan has 13 S&T agreements on biotechnology with the USA, France, FRG, China, Australia, Canada, Korea, the USSR and other countries. In recent years, biotechnology has been adopted as one of the important strategies in most S&T agreements.

Personnel Training Program

(1) For researchers in the MAFF. (a) Biotechnology training program. (b) Gene engineering program.

(2) For researchers in prefectures and private companies. (a) National research institutes' visiting researcher program for 2–9 months. (b) Intensive training program on biotechnological topics.

Conclusions

The food industrial technology closely relates to our eating habits and it is difficult to conclude which one has affected another. It might be rather reasonable to consider that they have affected each other. This may apply to food industrial technology itself. From the viewpoint of the development of the food industry based on so-called seeds and needs, new techniques developed in the food industry (seeds) have changed our eating habits, while consumers' needs have promoted the technical development of the food industry (appendix).

It is needless to say that the food industrial technology toward the 21st century should be considered from both of these viewpoints. It is to be noted here how to join the seeds in the food industry with practical application and how to technically respond to the consumers' needs. It is necessary that the seeds accumulated in the industry are always put in order systematically so as to be able to respond to any external request. It is commonly observed that seeds which are unnecessary at some time become acceptable with a change in situation. It is further required to continuously pay attention to consumers' needs and to approach them not by merely extending existing techniques but by having an enlarged view. Anyway, it is not easy to breed seeds so as to immediately put them into practical application. The vague chase of needs would not always bring about good results. It is necessary for food R&D towards the 21st century to properly examine and evaluate the conventional process and results of R&D from the viewpoints of needs and seeds.

Appendix: Degree of Consumer Interest in Biotechnology in Japan (Dentsu Co., Ltd.)

According to a survey on consumers' interest in biotechnology including degrees of interest in techniques, research and fields, 45.4% of subjects are interested or somewhat interested in the technology while 36.7% of the subjects are not or scarcely interested, showing a half-and-half tendency. In contrast, subjects exceeding 50% answered that biotechnologial research should be promoted in the future. This ratio is remarkably high compared with the ratio of subjects giving a negative answer (5.7%). It seems, therefore, that consumers recognize the necessity of biotechnology R&D. Consumers' interest in the contents of biotechnology focus on familiar ones such as 'applicable to the treatment of diseases' (53.8%), 'promoting the production of food' (32.1%), 'developing a new energy source' (21.9%) and 'creating a new food product' (20.0%).

Takashi Sasaki, PhD, Division for Food Resources, National Food Research Intitute, 2–1–2, Kannondai, Yatabe, Tsukuba, Ibaraki 305 (Japan)

Lindroth S, Ryynänen SSI (eds): Food Technology in the Year 2000.
Bibl Nutr Dieta. Basel, Karger, 1990, No 47, pp 38–50

Center for Advanced Food Technology: A Successful Example of University-Industry Cooperation in Food Technology Development: Gearing for the Year 2000

Myron Solberg, Jack L. Rossen

Center for Advanced Food Technology, Rutgers, The State University of New Jersey, New Brunswick, N.J., USA

Introduction

Understanding the complexity of food materials requires interdisciplinary approaches. The discipline of Food Science, with its integration of Food Chemistry, Food Biology and Food Engineering was the recognition of food complexity. This approach served to generate and assemble a core of knowledge which is Food Science. Food Science has been the base for the modern supply with its shelf stability, convenience, safety, nutritiousness, and aesthetic quality.

The trends of the future center around greater convenience, improved quality and safety, and a new healthfulness. Shelf stable products need to be perceived as freshly prepared by a world-class chef, brought to and maintained at ideal eating temperature throughout consumption, tailored to meet the nutritional needs of individuals with a variety of health requirements, and not only be safe to eat but designed to increase human life expectancy with the absence of disease.

The knowledge needed to achieve the goals is not in the data bank. Information as fundamental as physical properties of food materials, intrinsic and extrinsic factors affecting chemical reaction, phase changes, state shifts and structure:function relationships is not available. Without these data quantification is impossible. Without quantification modeling and prediction is unachievable. The lofty goals presented above defy trial and error and empirical approaches. Through definitive modeling, prediction becomes possible and through prediction there is an opportunity for some success.

Obtaining the quantitative capability calls for a new integration of discipline knowledge and the development of an enlarged Food Science knowledge base.

The coordinated efforts of industry, academe, and government could be directed towards solving the problem. The Center for Advanced Food Technology (CAFT) at Rutgers is one successful model. The remainder of this presentation will describe the CAFT model.

The CAFT Model

The CAFT model is designed to assure that new knowledge in the traditional disciplines of chemistry, biology and engineering is transcribed into new food science. This is achieved by bringing the traditional scientists into intimate contact with the food scientists through the establishment of interactive research teams where both groups are engaged in research with a common objective. The CAFT model does not remove the scientists or engineers from the peer group which creates the disciplinary critical mass and assures continued challenges upon the discipline knowledge frontiers, but rather interacts the various discipline representatives on a regular basis and focuses their energies towards gaining understanding of phenomena.

CAFT's Uniqueness

From the very beginning, CAFT has been unique. Initiated with State of New Jersey funds rather than US Federal Government funds, one would expect parochialism. Instead there was the State's confidence, through the New Jersey Commission on Science and Technology, that industry would be attracted by the science and technology expertise which would be created and concentrated in Rutgers, the State University of New Jersey.

Short-term research and development and payback would be anticipated as desires of the nontechnology/market-driven food industry. The industry participants demanded that the research be precompetitive, basic and seeking molecular level understanding. Applications and development were to remain the task of the individual companies.

Independent research is the tradition of the university professor and academic freedom to choose the subject and direction of that research is fundamental. The faculty viewed the concepts and accepted the challenge of interdisciplinary research with its potential broadening aspects as well as managed research which provided opportunities to apply expertise gained in one field of endeavor to new and complex problems in another field. Independence and academic freedom were maintained within the planned operational design.

Thus, three potentially unlikely partners, each still seeking individual goals, came together to create a new entity, CAFT. The CAFT mission fits the needs of the three partners.

CAFT's Mission

Conducting precompetitive, basic, generic, and interdisciplinary research on problems identified by industrial participants as key to the short- and longer-term needs of food processing and product development is the mission component filling industry's needs. Supporting high-quality research with funds for faculty, graduate students, postdoctoral fellows and skilled staff scientists; supporting the acquisition of state-of-the-art instrumentation, equipment and physical facilities to assist university research and education; and increasing the scientific visibility of the university nationally and internationally through research publications, programs, grants and gifts fill the needs of the University. Transferring knowledge and technology to member companies and being a visible source of information and assistance to nonmember companies and entrepreneurs contribute to New Jersey's economic well-being. Providing educated and trained employees for the State's food industry companies, providing educational opportunities to nearby industrial employees, providing opportunities for in-state industry to license new technology, and making experimental facilities available for early stage product and process development are components of filling New Jersey's expectations.

The CAFT mission has remained constant. Growth of CAFT has created balance between the three mission components. The key events time line presented in figure 1 illustrates the early concentration on basic research program development (1985–1987), followed by industrial membership activities (1986–1988), and advanced technology and support facility developments (1988–1989).

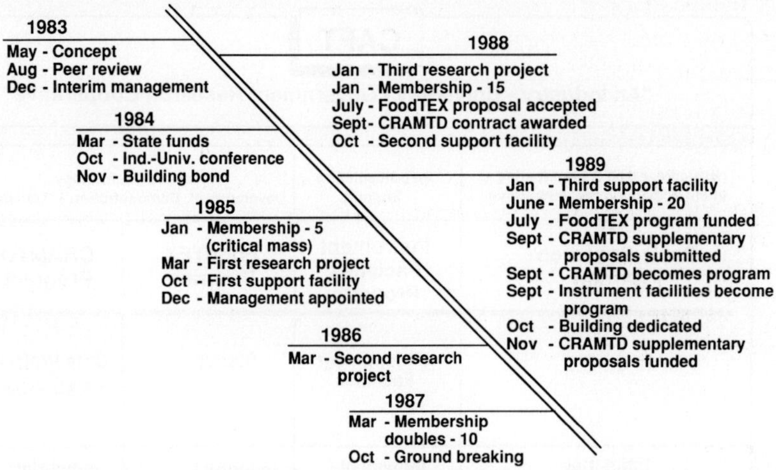

Fig. 1. Key events time line in the development of the CAFT.

CAFT's Philosophy

The CAFT philosophy has also remained constant. CAFT is dedicated to achieving its three-part mission by fostering an interactive environment which brings together not only academic scientists and technologists from a variety of disciplines but also industrial scientists and technologists from many companies. Through such interaction, increased understanding and respect is gained by the involved parties and team efforts are fostered. The close and frequent contact provides opportunities for academics to learn why certain information is needed and for industrial scientists to greatly reduce the time from information discovery to application. The support facilities program follows the more traditional mode of one company interacting with one researcher. The basic research and advanced technology programs follow the original model developed by CAFT management which establishes Industry-University interactive teams to plan, execute, and evaluate the efforts and accomplishments.

CAFT's Structure

CAFT today is presented in figure 2. From the basic research program which had incorporated within it the instrumentation support laboratories

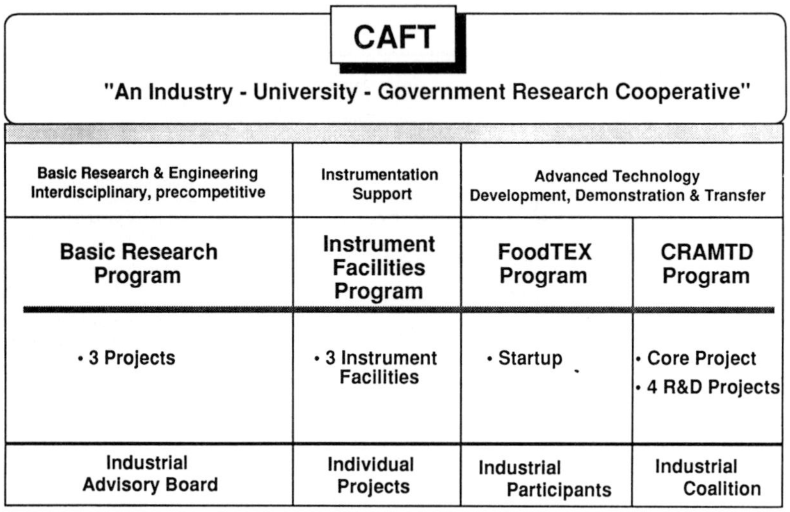

Fig. 2. The structure of CAFT, 1990.

and a technology transfer mechanism directed towards member companies, CAFT has become an umbrella organization which houses the basic research program with its industrial members, the instrumentation support facilities program with its supporters from university and government laboratories and member and nonmember companies, and the advanced technology development, demonstration and transfer program consisting of FoodTEX which is directed towards small- and medium-sized companies in New Jersey and CRAMTD with its industrial coalition of 19 companies and focus upon advanced manufacturing technology research and development for shelf stable foods.

Each of the programs is identified separately for budgeting purposes. The advanced technology programs were established with new funds so that they did not detract from the ongoing basic research program. Each program operates with funds from the State, the Industry, and the University.

CAFT's Organization

The organizational chart (fig. 3) provides a summary of the activities and serves as an introduction to the details which underlie the CAFT system. The academic advisory board consists of the chairmen of the ten academic

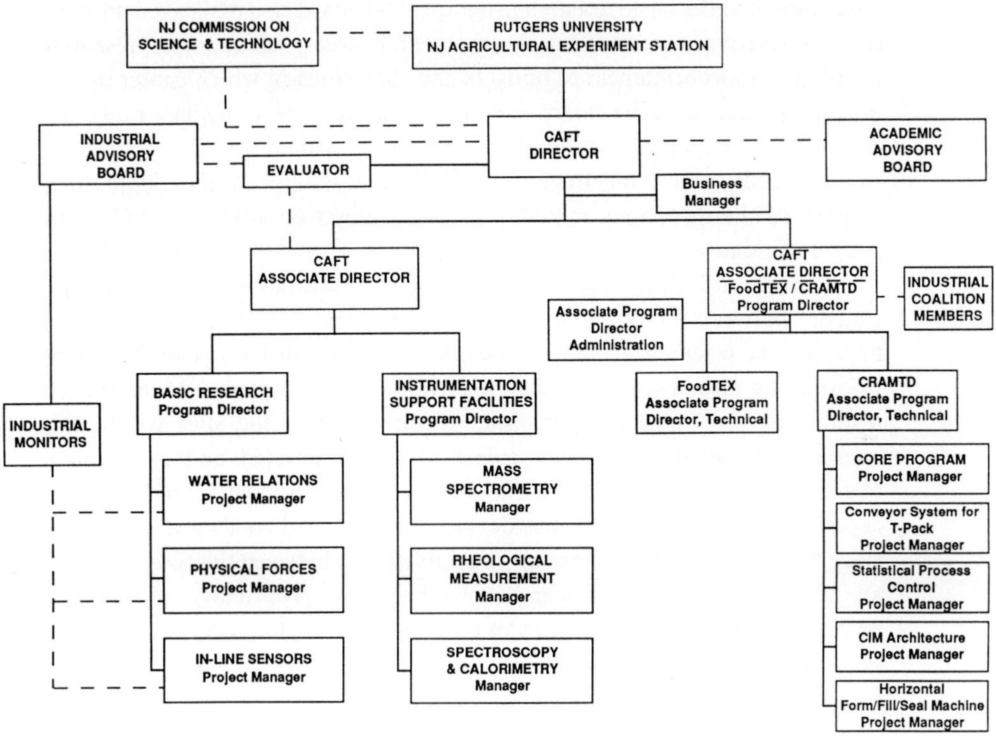

Fig. 3. Organizational chart of the CAFT.

discipline departments from which CAFT draws research faculty. The board provides guidance to the director relative to the quality of the research proposals and adequacy of budgets. As department chairmen, the board members provide assurance of departmental support to the faculty participating in CAFT projects.

The evaluator is a consultant to the director who is assigned the task of determining the level of satisfaction or dissatisfaction with CAFT among the industrial supporters and quantifying the benefits derived by the member companies. The information is obtained by surveys and direct discussions. The evaluator summarizes the information, renders it anonymous and reports it to the director.

The industrial advisory board is composed of one senior level representative from each member company. The function of the board is to assist the director in establishing policy relative to operation of the center, specifically with respect to the basic research program. The board also makes recommendations relative to identification of research areas, adequacy of research proposals, appropriateness of budgets and directions in which center development progresses. The board members also appoint industrial monitors from their companies. As many as three may attend any basic research project study group meetings and provide guidance to the researchers relative to the directions in which research might be advanced. Monitors provide guidance in written form through the evaluator on a quarterly basis.

The monitors are the key to successful information and technology transfer between the researchers and the companies. The monitors get the first look at research accomplishments and they must translate this new information into possible applications within their companies. Monitors are responsible for transferring translated knowledge to the sites within the company where development projects may be initiated and proprietary benefits may begin to accrue. Should a member company not have the capability or desire to carry the development needed to a more advanced stage, CAFT will serve as a broker to establish ties between the company and an individual or group of researchers so that a semiproprietary contract can be negotiated and development can be progressed within a department and separate from CAFT's basic research activities.

The Integrated Project Team

To complete the picture of CAFT's basic research program there is a need to elaborate on the development of a project using the 'Integrated Project Team' (IPT) approach and the mechanisms which encourage information transfer, the 'study group' and the semi-annual accomplishments reports.

The IPT approach was described in detail by Leslie et al. [1] in 1988 and Rossen and Solberg [2] in 1987. The process consists, in brief, of the following steps. CAFT management's acceptance of an identified research area and development of a brief scope statement in conjunction with the industrial advisory board. Submission of the scope statement to the industrial and academic advisory boards for identification of appropriate expertise within companies and departments, CAFT management selection of an appropriate project manager and setting up an initial planning meeting for persons previously identified by the two advisory boards. The planning meeting identifies project objectives and discusses approaches to be taken.

The next step is the organization of a pre-study group meeting which occurs after sufficient time for shifts in individuals with expertise as required by the objectives. At the pre-study group meeting the academics present brief overviews of how their expertise will meet the objectives and the industry representatives provide feedback relative to their estimate of adequacy of the expertise and the approach. The project manager reviews all input and issues requests for proposal component preparation to appropriate academics. Submitted proposal components are integrated by the project manager, sent to the potential academic research leaders and the potential industrial monitors for review. An integrated project proposal review meeting is set up and the potential research leaders present justification for their part of the total effort while the potential monitors provide feedback relative to the presentations and discussions. The project manager makes appropriate revisions in the proposal and after consultation with the industrial monitors and research leaders submits the proposals to the advisory boards for review and recommendations. CAFT director's approval of the proposal and funding for implementation initiates the official study group process.

The procedures described provide several key benefits. The series of meetings, the discussions, the feedbacks and the revisions establish the team which then continues to interact. Barriers are broken down, trust is initiated, respect and familiarity are established and a sense of a common goal is recognized.

Quarterly formal feedback from monitors helps to direct the course of the project and provide information for incorporation into the next proposal revision.

The project proposal is revised annually by a less rigorous process which starts with submission of research components and then follows the established course.

Each project manager publishes a formal semi-annual Research Accomplishments Report which is distributed to member companies on a confidential basis. Tables and figures used in quarterly meetings are distributed to the monitors. Monitors are encouraged to visit directly with research leaders who are carrying out research of specific interest and to bring proprietary samples for evaluation when such activity is warranted. Every 6 months, in conjunction with the industrial advisory board meeting, one project is overviewed in a 45-min videocassette format, a copy of which is provided to each member company for in-house use on a confidential basis. Brief overviews of the other projects are presented to the board members at the meeting.

Active Projects

The three active projects are entitled water relations in food systems, physical forces in food systems, and in-line sensors for moisture and rheological properties. The water relations project has 31 people involved in its activities and is supported with US $374,000 this year. The physical forces research effort has 57 affiliated personnel and an annual budget of US $829,000. In-line sensors has 33 people involved with a budget of US $541,000 for the current year. Each project represents a significant effort with a potential for generating scientific breakthroughs not possible in the traditional one funder:one researcher type project.

Membership

Membership in the CAFT basic research program has been at a cost of US $30,000 per year since 1985 and will increase to US $35,000 per year on January 1, 1990. A member company with an interest in only the smallest project is getting a leveraging of 12 to 1 on their membership fee. Interest in all three projects raises the leverage factor to 60:1. Taking into account that the project dollar values used do not include research administrative support or overhead charges, the true leveraging value would be better than 90:1.

Member firms may have the rights to free but nonexclusive license to patents, copyrights and other intellectual property resulting from the research and will share in royalties from a patent if it exceeds 20 times the membership fee in any year. Delay of publication in the open literature may be requested by member companies for up to 1 year for the purpose of securing intellectual property protection.

The basic research program is the cornerstone of CAFT's operation and has provided the springboard for the development and success of its other programs.

Instrumentation Support Facilities

The instrumentation support facilities program was initiated as an instrument upgrading process to support the basic research effort of CAFT and the University. The philosophical uniqueness of the system was the decision by CAFT management to place senior staff scientists in charge of the instruments rather than the traditional assignment of instruments to faculty. The chosen approach insures the costly instruments of full-time attention and concern. Additionally, the primary mission, support for research, takes first priority on the specialist's list of goals. Individual research is encouraged but only if it is maintained at a low level of effort.

The instrumentation support facilities include mass spectrometry, rheological measurements and spectroscopy and calorimetry. The three facilities are now managed by PhD or equivalent specialists. Their mission is to provide research support to CAFT, university, member company, government and nonmember company researchers, in the priority listed. Fees for service are charged and the facilities are challenged to meet their operating costs. CAFT researchers and university researchers are able to apply for partial to full grants of service fee remission.

The strength of the facilities lies in the expertise of the specialists and the quality of the instruments. Extremely competent and extraordinarily skilled operators and state-of-the-art instruments assure success. The facilities staff guide research, assist with proposal writing, and train academic and industrial students.

Routine industrial problems are not accepted. Lists of consulting laboratories are provided for those seeking such services. The facilities deal with research problems where the company expertise or instrumentation is inadequate to provide the necessary information.

More than 30 companies and more than 40 faculty members have utilized the support facilities.

Technological advance in instrumentation is rapid. Million dollar mass spectrometers of 4 or 5 years ago are outdated today. To reach the current forefront, an instrument costing US $1.5 million is required. Research in the future will require centralized expensive instruments to an even greater extent than may be evident today.

Food rheology is an area of expanding knowledge in food systems today. Colors and flavors can be added but texture needs to be inherent and developed by processing. Textural response is a structure-function relationship. The spectroscopy and calorimetry facility provide the instruments for study of structure and structure change when either an optical or a thermal window exists in a system. The mass spectrometer is the ultimate in molecular identification lending credence to the structure studies and the function studies.

Thus, the three facilities, which grew out of need, have great complementary potential for solving many food materials problems.

Advanced Technology Design, Developmental Transfer

Advanced technology transfer is a measure of success in an Industry-University collaboration. This CAFT mission has remained constant but its implementation has expanded. Initially transfer activities were directed towards member companies. Activities were added with respect to assisting small New Jersey businesses in preparation of grant proposals to federal agencies for product feasibility studies. The need for a designed and coordinated activity became evident as the State of New Jersey sought a more visible return on investment than can be obtained from a basic research program.

The FoodTEX Program

A proposal for the establishment of a Food Technology Extension program (FoodTEX) was developed and start-up funding was provided by the State in mid-1989. The market was defined as the more than 750 small and medium-sized food processing companies in New Jersey. Following a CAFT fundamental principle, which calls for input from the potential users in identifying the directions for a program, activities during the start-up year will focus upon defining the needs and opportunities for FoodTEX. Surveys and interviews are being used to accomplish the goal.

As the program develops there may be major opportunities to introduce processing as a mechanism for adding value to crops. Some of the processing may utilize advanced methods such as cooking extrusion and this is the place where new process techniques are implemented.

The short-term return may come from simple sharpening up of operations using well-established methods and equipment which may provide real advanced technology to companies using older technologies.

The CRAMTD Program

Another element of advanced technology transfer is the Combat Rations Advanced Manufacturing Technology Demonstration (CRAMTD) program. The mission of CRAMTD is to foster the use of advanced manufacturing technology in the thermostabilized food industry. To develop a flexible manufacturing system which handles both civilian and military products with minimal change over time and to develop computer-integrated manufacturing (CIM) capabilities for shelf-stable trays and pouches such that manufacturing costs are minimized and product quality is maximized are the objectives.

A state-of-the-art capability is being established as a baseline from which developments can proceed and be measured for their value. The baseline facility will be fully computer simulated and all advances will become a part of the simulation. Each significant development advance is structured as an individual multidisciplinary project. The four ongoing projects are development of an automated tray pack filling conveyor, a generic inspection statistical quality control system, a CIM architecture for food manufacturing and a horizontal filling form, fill, seal machine with tray/pouch flexibility.

The CRAMTD program was modeled after the CAFT basic research program. Rather than having an industrial advisory board and industrial monitors, it has an industrial coalition. Coalition members contribute knowledge so that the real state-of-the-art is known, they identify the problems which need solutions, they have an opportunity to receive subcontracts for design, development, and manufacturing of equipment and they have an open window on the developments from planning through completion.

Coalition membership is US $9,500 per year for companies with annual gross revenue above US $100 million and US $1,500 for smaller companies.

FoodTEX and CRAMTD can be looked upon as closely related and fully complementary. They are the applied component which makes CAFT basic research more complete. They are a natural evolution from the original program and are the place where new knowledge can be hired out and where additional need for new knowledge can emerge. They represent another step in shortening the time from discovery to application.

A Visionary Peek

As the year 2000 approaches, it seems clear that research leading to greater understanding of physical properties and then utilizing the knowledge in advancing process technology will still be needed. The 'chef-like', shelf-stable product individually tailored towards human life extension will still be the vision.

The realities will be many. The greater need for industry, university and government to share in supporting the research is obvious. The need to move from discovery to application more quickly becomes greater if all research supporters are to derive their objectives. Multiple use of government and industry funds so that research at universities will prepare students for the work force and provide capital for instruments and facilities. Increased interaction between industry, university and government so that each does only what it can do best.

The CAFT model is a first-stage development. It will be modified and improved many times and in many ways to fill the needs of the future. Its principles are these: interaction to define progress and resolve problems and establish respect and understanding; commitment in funds and people from all participating parties; undertake precompetitive research with multisponsor shared underwriting of expenses; maintain communications at a high level through all available techniques; and responsiveness to alter directions and emphasis when appropriate.

Acknowledgment

This is publication No. D10535-9-89 of the New Jersey Agricultural Experiment Station supported by State Funds and the Center for Advanced Food Technology. The Center for Advanced Food Technology is a New Jersey Commission on Science and Technology Center.

References

1 Leslie RB, Rossen JL, Solberg M: The Center for Advanced Food Technology. A model for industry-university-government collaboration. Indust Higher Educ 1988; 24:223–232.

2 Rossen JL, Solberg M: Rutgers University's Center for Advanced Food Technology. Food Technol 1987;41:100–102.

Prof. Myron Solberg, Center for Advanced Food Technology,
Cook College, Rutgers, The State University, PO Box 231,
New Brunswick, NJ 08903 (USA)

Lindroth S, Ryynänen SSI (eds): Food Technology in the Year 2000.
Bibl Nutr Dieta. Basel, Karger, 1990, No 47, pp 51–66

The Challenge of the Year 2000 for Food Technological Research in an Industry-Based Research Association – The UK Vision

David Arthey

Campden Food and Drink Research Association, Chipping Campden, Gloucestershire, UK

Introduction

Good company management practice demands that senior executives consider the long-term future pattern of growth and activity of their businesses on a regular basis. Experience shows, however, that whilst the activity of producing such a report is a good management tool, the contents and predictions are soon out of date. Unforeseen factors come into play which often radically affect the direction in which the business is going and unpredicted decisions must be taken on a day-to-day basis in the light of current pressures to ensure the continued success and favourable growth of the company.

The Campden Food and Drink Research Association (CFDRA) is a commercial enterprise which faces these precise problems in the growth pattern of its business which is primarily research and development. Like other forward-looking companies, it produces long-term predictions and, also like other companies, such reports are often out of date soon after they have been prepared. That, of course, does not deny the value of preparing them. Sources of research funding, particularly from government, are becoming more precarious from year to year; topical issues from consumer lobbies may cause unforeseen demands on resources which result in changes in direction of research and development programmes, and new emerging technologies and their associated problems in industry may demand that resources be transferred to deal with them. Politics must not be left out of the equation because government parties change and their leaderships usually have entirely different views to their opposition colleagues. The current influence of green issues may or may not have an important bearing in the future but food issues will certainly be implicated.

The title of this paper demands an element of prediction of where the food industry will be in the year 2000 but from the foregoing, this cannot be accurately assessed. It is not possible to take that leap of 10 years and clearly visualise the situation in which the food industry will find itself at that time. What can be done, however, is to look at today's position and current trends and try to predict from them a hazy picture of the future, bearing in mind the dangers of extrapolation in such circumstances.

UK Food Research Centres

Food science and technology research in the UK today is spread over a number of organisations whose inputs differ according to their policies, remits, responsibilities and expertise. The UK government has defined different categories of research as: basic, which is undertaken to acquire new knowledge without any application of use in view; applied, which is to acquire new knowledge with practical aims or objectives; experimental, which draws on existing knowledge to produce new materials, products or devices or improve those already in being; and strategic, which is applied research which has not advanced to its use in application. Each research centre will tend to predominate in one of these areas. Recently, government has introduced the terms 'near-market' and 'far-market' when referring to research, believing that 'near-market' research should be undertaken or funded by industry because it has direct implications and benefits to individual companies and their markets. 'Far-market' research is expected to have more fundamental long term pre-competitive benefits to the industry and is often in the form of providing underpinning science. As 'far-market' work progresses it may approach areas which could be considered to give commercial benefits and at that point industry is invited to continue its funding.

Some of the basic and strategic research is currently carried out at the Universities, notably Leeds, Surrey, Warwick, Reading, and Cardiff. Such work may often be undertaken by post-graduate students with considerable academic support from full-time University staff. Some curricula permit the positioning of students in industry or other research centres such as the Research Associations to carry out part of their research programme.

Basic and strategic research is also undertaken by the UK Institute for Food Research. In the area of food there were three laboratories: at Bristol,

Norwich and Reading, but the first is due to close and its relevant work transferred to the other two. Whilst research at these centres is primarily basic and strategic, some more applied activities have recently crept in due to the requirement by government that a proportion of their work must be funded by industry. In addition to the Institutes for Food Research, there are other similar Institutes in the research areas of Animal Health, Animal Physiology and Genetics, Grassland and Animal Production, Arable Crops, Horticulture and Plant Science, and Engineering where the impact of their research on food cannot be denied. All these institutes are currently supported by the Agricultural and Food Research Council; however, changes in this administration are now being considered. Proposals are afoot to unite the research and management efforts of all the UK Research Councils into one National Research Council, partly to encourage work of a more multidisciplinary nature across the Councils and also to effect greater efficiency in deployment of the available research resources.

Some research of a more applied nature is undertaken with government funds (Ministry of Agriculture, Fisheries and Food, MAFF) at Torry Research Station in Scotland and the National Institute of Agricultural Botany. Work of an even more practical and development nature is done by a series of Experimental Horticulture Stations and Experimental Husbandry Farms and these too are supported by MAFF although in the future some are scheduled to close and an increasing part of the remainder's work will have to be funded by industry.

The Food Research Associations, of which there are three: The British Food Manufacturing Industries RA at Leatherhead, The British Flour Milling and Baking RA at Chorleywood, and CFDRA at Chipping Campden, are commercial enterprises which, although they carry out a small amount of strategic food science research, primarily work in the area of applied and technological research and provide a significant service to their members. Each receives government support from a number of departments, chiefly the Chief Scientists Group within MAFF.

In addition, there is one further sector which undertakes food research and development. This is the private sector where there are not only commercial research laboratories which market their services to the food industries but also those laboratories which are run by major food and food packaging companies such as RHM, Unilever, Beecham Foods and CMB Packaging. Much of their work is confidential to the companies concerned although occasionally they secure government contracts for work the results of which are made public after a few years.

This paper addresses the question of technological research in an industry-based Research Association and whilst it is possible to paint the overall scene of food R&D in the UK, more detail is required to understand the background function and operation of a food and drink RA.

The Campden Food and Drink Research Association

The original research centre at Chipping Campden was first established in 1919 as part of the University of Bristol. In those early days the work was aligned to the commercial development of the canning industry. Campden provided much of the science base on which the industry was built and this was largely responsible for the high reputation which Campden achieved. Work on freezing began in the 1930s and continued after the Second World War when freezing as a commercial processing method became established at the end of the 1940s. In 1952, Campden severed its connection with the University of Bristol and became an industrial Research Association. RAs were established at that time to ensure that different sectors of UK industry were making contributions to research which would be to their direct benefit. Campden raised monies from industry through membership of the newly formed Association to which was added an equal sum from government. The UK government continued to match industrial income in this manner until the early 1970s when Lord Rothschild produced his report recommending that government no longer give finance to the RAs on a pound-to-pound basis but rather on a contractural basis. From that time government commissioned work at the RAs and paid totally for the programmes it supported. At Campden, this made little difference in the total financial support from government; it only changed the way in which monies were received. During this period the RA changed its name several times as its work became more embracing, culminating in its present title which was established in 1988.

Today the Campden Food and Drink Research Association is an international research body with expertise in food and drink preservation. Its primary aims are to research and provide services, including training, in processing efficiency, product safety and product quality through a centre of expertise with trained personnel, relevant resources and a high degree of experience and knowledge.

Membership of the RA is growing rapidly with an increasing percentage of companies from countries other than the UK. Members may be food processors, machinery manufacturers, packaging manufacturers, ingredient suppliers, raw material producers, agrochemical companies, importers, distributors, retailers, caterers and government departments. Income from members' subscriptions is used to provide and maintain services and to carry out research projects. Further research is commissioned by government, primarily but not exclusively, through the Ministry of Agriculture, Fisheries and Food. A recent innovation is the introduction of the LINK scheme where the Department of Trade and Industry together with MAFF provide funds for approved programmes of pre-competitive research which has also to be supported by two or more food or associated companies. A further source of RA income is contract fees where work is done for individual companies or a group of companies on a confidential basis. With the prospect of decreasing financial support from government in future years and the requirement for a greater input from industry, the RA has set up a number of Research Groups (Clubs) where a consortium of member companies have been brought together to finance research in particular areas of common interest to the Group. The Group pays totally for the research the results of which remain confidential to its members for three years enabling them to take economic advantage of the work. Some of the Groups currently in operation are: Sensory Evaluation Working Group, Agricultural Research Group, Heat Sterilisable Plastics Research Group and NIRS Low Level Analytes Research Group. Income is also generated from a rapidly increasing portfolio of training courses, symposia and conferences.

Under the control of the Director-General are six Divisions, two of which are commercial and four scientific. Within each Division are Departments which range from Agriculture through Chemistry, Microbiology, Sensory Quality and Food Acceptability, Food Process Engineering, Product and Packaging Technology, to Food Hygiene.

One reason for the considerable success of the Campden RA is its structure and operation of Technical Advisory Panels of which there are nine: Heat Processed Foods, Frozen Foods, Chilled Foods, Foods Preserved by other Methods, Agriculture, Quality Assurance, Food Science, Microbiology and Process Engineering. Only members are permitted to attend these meetings and they discuss the industrial research needs of their particular disciplines and allot priorities to a range of research proposals. In the light of resources and available expertise, those proposals receiving high priority ratings are considered for inclusion in the following year's programme. In this way the RA ensures that its research programme is directly related to the needs of its members. Members can feel confident that the RA is making a proper and relevant contribution to food science and technology and the advancement of the interests of the food industry. The current research programme lists 62 projects ranging from strategic and applied studies to the preparation of guidelines.

Current Research at CFDRA and the Challenge of 2000

Processes

In the year 2000 it is more than likely that the food products which are offered to us will be the same basic staple foods that we have on our supermarket shelves today. But many of them may well be different in terms of the forms in which they come, the technologies by which they have been prepared, the packaging materials in which they are wrapped or contained, the scientific techniques which have been deployed to ensure their safety, the additives which may have been used to enhance safety or quality, the controls over handling, distribution and retail display and many other factors which have been employed to make sure they are acceptable to us, the consumers.

In the past we have principally had processes such as drying, freezing, pickling and thermal processing in glass and metal containers. Today we have in addition chilling, MAP and aseptic packaging which often give advantages in quality over similar food packed by more conventional means. Now we may have the use of irradiation approved by government in the near future and which will probably be used for certain products such as dried herbs and spices.

Irradiation. At present, there is a fierce lobby against the need for irradiation and it will take time for its use to become accepted; indeed, some would say that this will never occur. But if food poisoning becomes significantly more of a public issue in the 1990s, then irradiation might be seen in a better light and become more readily accepted. This could be especially so with the removal of preservatives whose use is expected to decline in certain foods, although it has been suggested that the food industry might begin to add certain ingredients back to foods again to ensure a greater degree of safety.

Aseptic Packaging. Although aseptic packaging has now been with us for a number of years, the technology is still advancing rapidly and tends to leap forward with either a new development in packaging or with a new processing or filling technique. The recent introduction of ohmic heating enables particulates to be sterilised but both particles and carrying sauce or liquor are still treated in the same heating tube. The time may come when the different ingredients of a full meal are separately sterilised to optimise their individual sensory characteristics and are then brought together by multiple aseptic filling heads in a single, perhaps compartmented, tray and aseptically closed. We should expect to see additional progress in this area as we approach 2000.

The complexity of the aseptic system must not be overlooked. In-line monitoring and process control systems require critical instrument accuracy, and fully integrated computer control systems are indispensible. New methods for the establishment of heat processes such as the Campden alginate particle technique will be increasingly used as will hazard analysis critical control point (HACCP) as part of a quality management system. The RA's current responsibility in such an area is not to develop the engineering technology of the processing equipment but to provide the research and development work concurrently to enable food products of a high consistent quality to be produced safely for the benefit of the consumer. Campden has been involved with the development of HACCP analysis, the alginate particle technique for establishing heat processes and the production of guidelines for good manufacturing practices. It has aseptic packaging systems on site on which research and development work can be undertaken and this equipment is available for use by the RA's members and clients. Campden will continue to be involved with the development of this process and by the turn of the century much of the scientific knowledge surrounding aseptic packing will be vested in the RA.

Microwaves. Aseptic foods of the future will have to take into consideration the growth in home-ownership of microwave cookers. For aseptic foods, such containers will have to have high barrier properties to ensure a long shelf life. But microwaves are becoming important for their industrial potential as well as domestic use. Already several pasteurisation processes using microwaves have been developed to give short shelf life products such as pastas, which are retailed out of chilled food cabinets. Such methods have been developed in Sweden, The Netherlands and Italy and the companies involved are now seeking to enhance their techniques to enable long shelf lives at ambient for foods of the future.

Campden has an increasing involvement in the developing area of microwave technology. Currently, programmes are being set up to develop process control strategies for the production of pasteurised and sterilised products using microwave energy as the heating mode. The work will include the determination of the critical factors concerned with the operation of microwave heating systems, and how they should be monitored and controlled to ensure process integrity. Again the RA's involvement in this field will ensure that, like aseptic packaging, this is a technology in which Campden will be deeply involved at the turn of the century.

Chilled Foods. It is difficult to say what proportion of Campden's work is currently involved with chilled foods, but it can be said that it makes an important and vital contribution to the state of our knowledge of this technology. It has been estimated that the consumption of chilled foods will nearly double by the year 2000 whilst frozen foods will increase by 25% and canned foods decline by 25%. Chilled foods already form a significant part of our diet and are projected to become more important in the next 10 years. But chilled foods including cook-chilled foods in catering situations are not without their problems as the recent outbreaks of food poisoning incidents have shown. There are particular concerns over processes such as sous-vide. Chilled foods require very careful control in their production, handling, distribution and retail display to ensure their safety, quality and acceptability. Not only does the RA have clearly defined research programmes in this area, particularly in microbiology, but it is also actively involved in leading the team that is currently revising the Institute of Food Science and Technology (IFST) Guidelines for the Handling of Chilled Foods. It also has the responsibility for organising a consortium of major European companies to monitor time-temperature data in chill chains and the production of guidelines and education programmes for all aspects of the chilled-food industry plus schools and consumers.

Although chilled foods will form a major part of the food industry scene at the turn of the century there will continue to be occasional problems over safety. Quality will be of great importance; production control will be very sophisticated and plant personnel will be more educated in the essential control procedures. Already, Campden has set up a specialised laboratory to deal with the many aspects of food hygiene and sanitation of manufacturing premises and demand for information in this area is growing rapidly. The RA is currently building a new food processing hall which will be especially equipped to deal with production tests on chilled foods and this will be opened in the spring of 1990. Campden is and will increasingly be recognised as the major UK research centre in this field.

Combination Technology. Combination technology is a processing concept which was propounded in 1976 although the technology was in use long before that date. Some chilled horticultural products could be considered under this heading such as those packed in a microclimate of added gas or a mixture of gases. Modified atmosphere packaging currently employs nitrogen, carbon dioxide and oxygen either singly or in mixtures according to both the crop species or the style of presentation (i.e. sliced, shredded or diced) together with chill temperatures to extend the shelf life of the product. An important aspect is to ensure that the correct packaging material is employed to allow gases to permeate through the film in such proportions that the residual mixture during continued respiration is such that maximum shelf life together with optimum quality is retained. Only recently has the technique begun to win acclaim from the industry and there is still much work to be done to provide films with the correct permeability for each fruit or vegetable.

Combination technology extends beyond the mixture of chilling and gas packing and includes a range of antimicrobial chemical treatments such as the use of acetic acid, lactic acid, salt, sugar, sorbic acid and nitrate. The effects, in combination with physical parameters such as drying, heating, chilling, irradiation and filtration, are many and depending on the precise combination used may be advantageously synergistic or additive or disadvantageously antagonistic.

Packaging

Foodstuffs cannot be disassociated from the containers in which they are packed or the films in which they are wrapped. Each processing technology or manner of manufacture has its own particular needs and when the method of preparation or consumption of the food is taken into account the permutations and combinations of materials is enormous. It has been said that prior to the First World War, packaging was agrarian, that is the use of leaves, wood, cotton and other similar materials. Between the two world wars the era was industrial with the use of packaging materials such as glass, tinplate, aluminium and paperboard. Between 1951 and 1974, the development of plastics, high barrier films, coextrusion and injection moulding, flexible pouches and retort pouches, ovenable paperboard and composites took place. After that period microwaveable materials, high barrier plastics and specifically permeable films have become important.

The metal can has been with us for many years. In the future there will be cost reduction enabling the use of thinner materials without affecting the resistance of the container. There will be more strength, tamper evidence and reliability to give greater safety, and there will be convenience, for example, improved stackability, easy open ends using heat-sealed lids, and microwaveability of metal containers such as trays. Today some metal containers are being replaced by laminated plastics and, of course, high barrier to give a commensurate shelf life. But it has been said that plastics belong to the microwave both now and in the future and there is now a wide range of materials which can be used for microwaveable products. There are also advantages with the use of glass for heat-processed foods and these are said to be transparency, quality, shape, integrity, chemical resistance, high barrier properties, strength, colour, microwaveability and thermal properties.

Campden has an integral involvement with all types of containers and whilst it does no direct development work with containers, it does work closely with container manufacturers to ensure correct closing, suitable heat processing of specific foods and appropriate cooling with the best process hygiene. The formation and success of the Heat Sterilisable Plastics Research Group is a good example of how the RA can work with its members in this area. Containers of the future will be more innovative and sophisticated and Campden's role in helping to ensure safety will become ever more important.

Process Control and Sensors

The increasing sophistication of processes such as aseptic packaging and the application of computer control systems is almost essential in the newer technologies to ensure process safety, consistency in quality, and efficiency. Campden has been instrumental in developing and installing computer control systems in, for example, a canning factory. It has also commenced work on a study of the physical properties of foods to provide underpinning data for the design and specification of refrigeration systems. Such work, together with studies on heat transfer coefficients, will help to ensure more efficient performance of food process systems in the future. As our understanding of these physical factors grows the industry will be able to maximise quality whilst at the same time operate at optimum efficiency. Such work will take on greater importance in the 1990s.

Developments in process control will be aided by success in the rapidly developing area of sensor technology. After treatment, information from sensors will enable real-time analysis of foods and thus improve product quality and safety. At present sensor technology can fall into non-invasive technologies, such as accoustics, fibre optics, nuclear magnetic resonance and near infra-red spectroscopy (NIRS); and invasive such as enzyme and antibody biosensors and electrodes and fibre optics. It has been said that the development of food process control technology is impeded by a lack of in-line and on-line sensors and that so far most sensor technology development has been in the pharmaceutical, health care, military and electronics industries. But now there is evidence that the need for sensors in the food industry is beginning to gain importance. They should be on- or in-line, non-invasive and capable of analysing complex heterogeneous systems in unstable conditions, sometimes at high temperatures and fast flow rates. In many areas the basic work will be done elsewhere but Campden is involved already with both non-invasive and invasive sensors particularly with ultrasonics and NIRS and it has led the field in the application of the NIRS system to many requirements of the food industry. Associated with sensor technology are the opportunities for neural networks and radio frequency information transfer which will aid the collection and interpretation of data provided by the sensors. Sensor technology may develop well beyond the bounds of expectation by the year 2000 and who can tell the degree of sophistication it may have reached or the areas in which it may have become important by that time.

Biotechnology

It is forecast that biotechnology will have a significant impact on the future of the food industry, indeed, such effects are already being felt, for example, the commercial use of the microprotein 'Quorn'. Biotransformation such as that already mentioned is an example of the use of biotechnology directly in food situations. Other examples are the use of genetically engineered organisms to produce chemical compounds such as amino acids, vitamins, alcohol and enzymes, or to improve separation and purification techniques. The present requirement to submit all such novel products and processes to government for approval is likely to become mandatory in the future.

In agriculture, improved plant nutrition and better control of diseases without the need to apply agrochemicals are major research areas in biotechnology. Improvements and modifications in traditional methods of crop production will be achieved which will enhance the value of agricultural produce, reduce processing costs and enable the food industry to create new products. Some examples of these are: the production of seed with higher protein content and therefore higher nutritional value; wheat crops with improved milling and baking properties; oil crops with less saturated oils; plants that produce speciality chemicals; and plants that can be irrigated with salt water.

Although the RA is not directly involved with the fundamental research of genetic engineering and the recombination of genes from different organisms in one new organism, it is increasingly likely that it will be involved in the spin-offs from this work especially where there are major advantages and applicability to the food industry. Thus, the examination in simulated commercial situations, and indeed at actual commercial sites, of new compounds for a better understanding of their potential use might well be carried out by an RA. Similarly, genetically engineered crops for pesticide resistance, higher yield, or improved mechanical harvest ability, still need to be tested for their suitability for processing and subsequent level of product quality.

The era in which enzymes have been looked upon as having a negative effect on food and food quality is gradually passing. It is true that enzyme inactivation, particularly in crops for freezing and thermal processing, will continue as a necessary part of the process but a more positive view, likely to emerge in the next decade, is to use enzymes in such a manner as to effect beneficial biotransformation in foods and raw materials. Such biocatalysts can be used to upgrade oils and fats, nutritionally modify carbohydrates and synthesise high value food ingredients. Further enzymes may be used as catalysts to break down and destroy food spoilage and poisoning organisms and thereby form part of a combination process.

Again the RA at Campden has a long history of work in the field of enzymology and has investigated the role of enzymes in vegetables for freezing. The identification of lipoxygenase as a more important indicator enzyme for green beans than either catalase or peroxidase, together with the development of a practical factory test, illustrates the way in which an RA can undertake fundamental work and apply it to industrial situations for the benefit of quality in the final product. The RA is equally capable of looking at the more positive aspects of enzymology, bringing success in research to the practical needs of its members; it will undoubtedly attempt to move into this field in the next decade, possibly in the use of enzymes to advance sensor technology.

Healthy Eating and Safety

A recent campaign has as its aim to turn 20% of British agriculture over to organic production by the year 2000. By organic production is meant systems which are designed to produce optimum quantities of food of high nutritional quality by using management practices which aim to avoid the use of agrochemical inputs and which minimise damage to the environment and wild life. Efforts to introduce natural resistance to pests and diseases by genetic engineering will, of course, be of great benefit to such noble ambitions if they succeed but even if they do it is unlikely that such a radical change will take place. The figure is more likely to be between 5 and 7%. Nevertheless, the organic lobby has become much stronger in the UK than was originally thought possible and about 2% of horticultural produce is now offered in this form. The ideals are laudable but the consumer who constantly wants better quality often has the problem of accepting blemished material if it is to be organically produced. Campden has an involvement in this area of work and is anxious to determine whether sensory properties are affected by growing crops in the absence of inorganic compounds. By 2000 we shall have many more answers to questions surrounding this method of culture.

The consumption of organically produced foods is just one aspect of the very topical subject of healthy eating and in recent years consumers have changed in terms of their perception of the nutritional value of food and additives. The most visually effective change was the removal of tartrazine from orange-coloured drinks and its replacement by synthetic nature identical colouring. The use of such natural materials will increase during the next decade. Today fibre, carbohydrates, sugars, salt, fat and other food-associated compounds are all key words in terms of healthy eating, some being desirable, others to be avoided. They are associated with heart disease, cancer and general fattening. The trend towards fresh and chilled foods and away from more heavily processed foods is also an indication of the consumers' wish to eat more healthily. New regulations have been introduced for the use of pesticides and maximum residue levels have been established. Soon it will be necessary to name post-harvest applied pesticides on the labels of horticultural produce where they have been treated.

Water is gaining rapid interest at present with a forecasted increase in cost of 25% per annum for the next 5 years. This has partly come about from both the UK government's intention to privatize the industry at the same time that the EC is introducing strict new regulations over quality of water. In addition, water companies may no longer accept the concentration of effluent from food companies until quality is improved to comply with further EC regulations. This will put pressure on an industry which has paid scant attention in the past to the fact that water is a vital ingredient in their production processes. Campden RA has responded to this potential threat by taking the initiative of setting up a special working party to look at water quality, water usage in process situations, water conservation, recycling, alternative means of disinfection, and effluent treatment and control. Guidelines may be set up to deal with many of the current problems which are likely to persist well into the next decade.

Alongside the trend towards healthy eating and drinking is the need to ensure safety. As new foods and food processes are introduced so there is a need to ensure that their products have been manufactured or prepared in such a way they will be enjoyed and not suffered by the consumer. The RA's involvement in this area both now and in the future is considerable. Heat penetration tests are undertaken to ensure that correct processes are given to any particular recipe. Challenge testing can determine whether food poisoning microorganisms can grow and/or produce toxin under specific conditions. Predictive modelling of processes is an important area of food research and the RA has mathematical modelling programmes and experience in many areas of food processing including the control of process deviations. The hazard analysis critical control point (HACCP) concept has been advanced very considerably at the RA in relation to specific foods. HACCP is a particularly powerful tool which enables causes of concern along the process line to be identified and monitored during production thus preventing the manufacture of inferior units rather than using the more traditional quality control exercise at the end of the line when it is often too late to do anything about a particular batch. Such procedures will be adopted increasingly in the coming years and new automatic test methods will be introduced. The RA will remain in the forefront of not only the development of methods and the procedures for using them but also the training which is necessary to understand and use them effectively in industrial situations.

Sensory Evaluation and Food Acceptability

Rapid changes are taking place in the discipline of sensory evaluation and to it is being added the closely associated area of consumer research both in terms of acceptability and preference. By the turn of the century these subjects will have become much more readily accepted as true scientific research areas and no longer looked upon as desirable but rather unimportant contributors to our knowledge of food. Already the science of sensory and acceptability studies has its own unique methodology with procedures such as principle component analysis and generalised procrustes rotation together with highly advanced statistical techniques to analyse the data. Campden's role in this area is to be aware of new techniques as they are developed, to evaluate them for their benefit to the food industry and to deploy them in industrial client situations. The RA will also modify academic procedures to ensure that they can be used by the food industry to best effect and perhaps by 2000 even the marketing departments of food companies will come to realize the importance of a scientific input into these areas.

Consumer research at the RA is relatively new but already Campden has a 1,000 in-home panel and a team of consumer interviewers which has operated successfully with food and drink products. Such work will continue to develop rapidly and prove to be of increasing importance both in research terms and as a service to the industry. The RA has the advantage of possessing teams of highly trained sensory analysts who can examine a product in detail in the laboratory and whose findings can be compared with the results of consumer tests. This link between experts and attitudes of consumers can be extremely valuable in the development of new food products.

Technology Transfer

A vital role of any RA is technology transfer, that is, the utilisation of information of a fundamental nature, such as the results of research conducted by Universities or Research Institutes, and the translating of this into information which can be readily utilised by industry. RA's with their increasing involvement with academic research centres and their close involvement with industry through their membership are in a unique position to achieve this transfer efficiently and effectively. Transfer of information can be achieved in a number of ways such as training courses, the preparation of guidelines and codes of practice, development of computer software which would include expert systems and databases, and, finally, the all-important personal contact with industry which is an integral part of an RA's activities.

Training is scheduled to become a very important part of Campden's future activities and by 2000 the RA will be offering an even more comprehensive portfolio of training courses to meet every industrial need. They will range from person-to-person training, through 1-day 1-company courses and specific aspect courses, to longer-term training of 1 or more weeks' duration in specific technologies. There will also be an increased number of conferences and symposia dealing with subjects such as currect issues, new technologies and long-term forecasting.

Already the number of guidelines and codes of practice produced by the RA or with the RA's assistance is increasing year by year. This activity will develop during the 1990s to provide the guidance which the industry needs to ensure that foods are safe, wholesome and pleasurable to eat.

The RA operates a sophisticated and broad-based member service activity where the latest information from research and development from RA sources and elsewhere is deployed in answering the problems of members. Part of this service comes free to members and the RA can be called upon at short notice to deal with urgent matters. An immediate reaction service has been introduced to enable members to contact the RA at any time of the day or night to deal with urgent problems. Such services are unique and are based on a recognition that the RA is a technical extension of its members' businesses and, like them, ready to act to support them and help them secure the confidence of the public at all times in the quality, safety and reliability of the foods supplied to the consumer.

Conclusion

As foods and food processes become more sophisticated so the microbiological, chemical, physical, technical and sensorial qualities will require a deeper degree of understanding. Process control systems and manufacturing practices will become more complicated, measuring systems and sensors will need to work with the utmost accuracy, and manual operatives, even though they will be few in number in 2000, will need to be highly trained and fully aware of their responsibilities. Small errors will be capable of leading to major problems which, if allowed to occur, will lead to strong consumer lobbies and rebound significantly upon the reputation of the food industry. In this scenario, Campden will provide a strong supporting role for the food industries with a wide scientific and technical knowledge base to meet any situation. It will react speedily in any given situation to assist its members and at all times provide a developing role of advice and information to the industry to ensure its progressive and satisfactory growth.

David Arthey, PhD, Campden Food and Drink Research Association,
Lhipping Campden, Gloucestershire GB-C55 6LD (UK)

Lindroth S, Ryynänen SSI (eds): Food Technology in the Year 2000.
Bibl Nutr Dieta. Basel, Karger, 1990, No 47, pp 67–86

Necessary Actions of the Industry for Successful Food Technology in the Year 2000

Klaus Paulus

Bund für Lebensmittelrecht und Lebensmittelkunde e.V., Bonn, BRD

Introduction

So that the food industry can plan ahead and obtain maximum benefit from scientific and technological knowledge, it is important that the industry acquires as much information as possible on likely future trends. To be able to do this, first of all the major factors or elements have to be derived which may determine necessary actions of the industry for their successful future. Such an analysis may lead to a rather extensive list of elements, and in specific cases some of these elements may show different priorities. For a more general discussion it is meaningful to aggregate those elements to factors with an overall impact on the food industry.

Before doing this, it seems useful to have a look at the present situation and to describe the concerns of today in industrialized countries. One of the best-kept secrets of the past several decades is that people are extremely healthy and are becoming healthier and living longer every year. Life expectation is increasing, death rates from heart disease and stroke go down, infant mortality is decreasing, and most types of cancer are decreasing or leveling out. Although medical technology gets the credit for this development, an important impact has in fact come from food technology. The discovery and utilization of scientific principles in food technology have provided a supply of safe, wholesome, and nutritious foods heretofore unknown. This has removed the vital risk of starvation and most forms of malnutrition from the Western world and has left in their place the less vital but still important risks involved with more subtle diet-health relationships. But the public is concerned about many products of the food industry and the number of risks discussed in public is not identical with the real number of risks. Modern food processing is not well understood by the public. A major problem is the growing dichotomy between the public's dependence on technology and its ability to understand and tolerate new technologies. There is a fear of the unknown. Hostility to technical operations or to chemistry leads to the situation that industrially processed foods together with residues, contaminants and additives to foods are continuously attacked and are presented as most severe risks. The real risks such as overnutrition or malnutrition and the hygienic problems are almost disregarded. In the same sense risk/benefit determinations by the public are poorly informed and often unwise as a result of ineffective general education in science. Attitudes towards risks and benefits depend on whether these risks and benefits are vital or nonvital, voluntary or involuntary, individual or collective. It makes a big difference, for instance, whether somebody thinks he makes the decision or if it is forced upon him. Voluntary risks that are a thousand times greater than involuntary ones are generally accepted. Thus, the internal decision to eat much more than necessary is not regarded as a real risk but the contamination of a food with radioactive elements is a vital risk.

Technological developments respond to internal and external influences. Internal influences include the availability of research results and potential applications, whereas external influences include positive and negative demands and external influences are effected to a large degree by education, persuasion, and manipulation. Developments in food technology require especially that product acceptability, perceived safety, economy, legal requirements, nutritive quality and actual safety must be taken into consideration. The first three requirements are effectively judged by the consumers, but the other three cannot be. These requirements are somewhat like a measure whether or not an existing or newly developed technology will be successful. If the requirements are adequately met, there is no actual need for a change, otherwise modification or even replacement becomes more likely.

After this short analysis of the general concerns in the environment of food technology it is necessary to concentrate on some of the important factors with the already mentioned impact on the food industry. Four factors in this respect may play a conclusive role for necessary actions: (1) the consumer, as he is the customer and partner in the food industry; (2) science, as the basis for all developments within the food industry; (3) food legislation, as the borderline determining the possible actions; (4) food industry's policy in general and the resulting strategy for individual companies.

These four factors will be discussed in detail in further chapters of this presentation.

Consumer Attitudes and Nutrition Behavior: What the Food Industry Should Know

Food preferences and finally food acceptability depend on a great variety of factors determining consumer attitudes, consumption patterns and nutrition behavior. Acceptability is a dispositional structure, the tendency of an individual to accept a food as perceived or to reject it. Acceptability is a dynamic, restructurable, and polyadic relation. First of all it is a relationship between the food and the person. This relationship quite often changes from moment to moment, and the determinants for the momentary acceptability might be classified under three rough categories: (1) sensory characteristics of the food; (2) physiological state of the consumer; (3) social, institutional and motivational factors.

Food acceptability is the study of human food selection, it is a marriage, albeit an uncomfortable one, between food science and behavioral psychology. It is uncomfortable, as there are few food scientists who are expert in psychology and vice versa. And, regrettably, there have been relatively few food professionals who have even recognized the need for input from behavioral psychology, perhaps due to incorrect interpretation of food acceptability in terms of the physicochemical properties of food rather than as a behavioral phenomenon. The situation seems to change, some parts of the food industry as well as government agencies have declared their intention of interest in food acceptability. The fact that so very few really new products are ultimately successful on the market shows clearly that there is much to be done.

General Trends in Food Consumption Patterns in Europe

Quantitatively, European food demands over the next 10–20 years will be static or even fall due to only small increases in total population, increases in the proportion of elderly with lower food needs and decreased energy requirements of individuals. Under the influence of the 'healthy eating' trend, consumers in Northern Europe have reduced their consumption of red meat, full fat dairy products and some processed foods while increasing their intakes of whole-grain bread, and 'fresh' fruit and vegetables. A continuation of this trend, and that in Southern Europe towards more livestock products and processed foods, will stimulate a convergence of Northern and Southern diets. Consumer demand for diversity, quality, and healthiness makes demands on the direction and adaptiveness of research. Agriculture will have to develop the quality characteristics and variety that consumers seek.

Processors will have to find ways to reduce their use of fat, salt, sugar, and chemical additives without adversely affecting sensory properties. In this connection, agriculture and the food industry will have to work together more closely.

For the further trends some main factors have to be considered, such as relationship with income, demographic, cultural and social change, health and fashion and, of course, new choices through new technologies.

Some Forecasts for the Trends in Germany

Several studies carried out in the last years indicate that the most important trends can probably be summarized as follows: (1) there will be a demand for an even greater variety of food products; (2) the proportion of food products which are processed less intensively will increase; (3) foods of higher quality receive more interest; (4) convenience foods are still requested but they should be of high quality; (5) products better adapted to the technical standard of households will be needed; (6) development of products of high physiological value will meet the demands of special consumer groups.

These trends will be discussed in more detail in a following chapter.

In an investigation carried out recently with 2,000 consumers, it was shown that healthiness becomes more and more important with respect to nutrition behavior [1]. In the case of a decision between a healthy but not so pleasant meal and a pleasant but not so healthy meal 39% female and 26% male consumers would prefer the healthier one. Still 26% women and 40% men would select the more pleasant meal, but this preference is much lower compared with the corresponding results of an inquiry carried out 10 years ago. But these results do not prove a conscious nutrition behavior, as the judgements on which foods are healthy and which are not are based more on vague prejudices than on well-founded knowledge. More information is necessary therefore and this is a challenge for the food industry in a sense not to increase the uncertainty of the consumer with unrealistic declarations but to tell the truth in order to gain back the partly lost confidence.

This is confirmed by a decreasing interest in nutrition. In 1989 around 60% showed a positive interest. In 1979 the proportion was 75%. The main reason seems to be the manner of communication concerning questions related to nutrition. 55% say that the information is not instructive, 70% feel the information is contradictory, 45% do not understand the information, and almost 40% consider it narrow-minded. It seems interesting, furthermore, that the world nutrition is more related to the head or brain whereas the world eating is related to the belly of the consumer, indicating a different thinking between nutritionists and consumers.

In October 1989, a further Nestle study 'Nutrition and Creativity 2000' was published in connection with two earlier studies, of which perhaps that from 1985 on 'Man and nutrition 2000' is well known. One main result is that the consumer in the year 2000 will be creative. Eating will be an important aspect of life. For 70% of the German population eating is ranked third after living and personality. The traditional eating habits are being more and more replaced by a situative consumption pattern. This is a chance for the industry to meet the desires of these creative consumers of the future.

Progress in Science: Possibilities and Constraints as to Research and Development Activities

Traditional empirical technology consisted of drying, salting, cooling, fermentation and dry storage. Preservation by heating began with the work of Appert in around 1800. It was empirical for many decades before science began to catch up, for example with Liebig establishing food chemistry or Pasteur food microbiology. Around 100 years ago, modern food technology started in general terms as a factory operation based on the mechanization of traditional practice and the control of the results by chemical analysis. But the great leap forward did not occur until after the Second World War when the principle and importance of water activity was established, the application of kinetics to food processes began, the explosion of methodology and instrumentation occurred, a revolution in materials for plant and packages began, innovations in traditional processes could be developed to improve the quality, particularly taste and appearance of products and, finally, the introduction of food additives.

The ten top innovations in food technology during the last 50 years were presented during the IFT Congress of 1989 in Chicago. It could not have been easy for the jury to select the farthest-reaching innovations from a large number of developments, and perhaps the 'top ten' of the greatest innovations can only apply for the USA:

(1) Aseptic processing and packaging of liquid food.

(2) Vegetable preservation by sterilization processes with maximum safety and optimal product quality.

(3) Microwave application.

(4) Freeze concentration of fruit juices.

(5) Packaging and storage of fruits and vegetables under controlled or modified atmospheres.

(6) Freeze drying.

(7) Deep freezing.

(8) Optimization of the water activity in order to extend shelf life.

(9) Enrichment of food with vitamins and minerals to improve their nutritional value.

(10) UHT treatment of liquid foods, in particular milk.

Discussing these examples in detail would show the impact of developments in other areas of science and the necessity to apply the most relevant findings to the field of food technology. Food technology and industry therefore depend to a large degree on the progress in science and technique. They do offer new possibilities, as for instance the technology of extrusion cooking, but also create constraints, as for instance the tremendous possibilities in analytical chemistry, and thereby influence research and development activities within the food industry.

Impact of Science and Technology

In 1987 a report of the Commission of the European Communities on the subject of 'The impact of science and technology on the EEC food industry in the next 10–15 years' was published [2].

Using standard Delphi techniques, a panel of 63 experts representing 9 countries was interviewed. Some of these results will be presented, as they may determine necessary actions by the food industry.

The various forecasts of future developments can be distilled to give a projected scenario of probable events during the next 15 years. These events, together with their implications, are discussed below.

Over the next 15 years food science and technology will have great impact at the manufacturing, product and social levels, resulting in more efficient production of foodstuffs, offering improved levels of quality and 'freshness'. More efficient production will be achieved by a variety of techniques, with the most important contributions probably coming from increased levels of computerized automation and process control. Major improvements in process and equipment design and the more widespread use of inherently more efficient processing methods, e.g. microwave heating and aseptic processing, will also contribute to increases in production efficiency. The possible impact of processing methods by the year 2000 is shown in table 1. It was commented that the rate of consumer response to the processes listed in table 1 will influence the intensity and velocity of

Table 1. Impact of processing methods by the year 2000 (number of replies)

Method	None	Slight	Moderate	Major	Total replies
Aseptic processing	0	1	18	25	44
Irradiation	0	11	18	15	44
Gas/modified atmosphere packaging	2	2	22	18	44
Microwave/RF heating	0	3	21	20	44
Ultrafiltration/ reverse osmosis	0	10	17	17	44
Extrusion cooking	0	5	23	16	44

Table 2. Forecast % replacement of synthetic additives by natural additives

Year	Lower quartile	Median	Upper quartile	Total replies
1990	10.5	20.0	46.5	42
1995	24.0	37.0	59.5	42
2000	30.0	52.5	77.0	41

development. An example is food irradiation with its very different legal status throughout the EEC and the attitudes of the consumers to this method.

The net consequence of these developments is a forecast 25% decline in employment levels in the EEC food industry by the year 2000. Although it is difficult to quantify their impact, developments in science and technology will assist the forecast reduction in the use of certain food additives, notably synthetic colors and flavors (excluding synthesized nature-identical materials) and preservatives. The probable reduction is shown in table 2. The bulk of this replacement will relate to synthetic colors and flavors, as listed in table 3. The likely development of a new range of preservation techniques will fuel reduction in the use of preservatives in particular.

Despite the current uncertainty surrounding irradiation, results from this study indicate that it will have a medium to major impact by the year 2000, particularly in the areas of microbiological control, sterilization of packaging material, increasing shelf-life of fresh produce, and pest control in commodities and raw materials. If adopted on a widespread scale, irradia-

Table 3. Forecast changes in the use of food additives in the EEC food industry by the year 2000 (100 represents no change, 200 represents a doubling)

Additive	Lower quartile	Median	Upper quartile	Total replies
Natural colors, including synthesized nature-identical colors	150	200	285	41
Synthetic colors, excluding synthesized nature-identical colors	25	40	80	41
Natural flavors, including synthesized nature-identical flavors	135	185	273	42
Synthetic flavors, excluding synthesized nature-identical flavors	30	60	85	42
Emulsifiers and stabilizers	90	104	130	42
Flavor enhancers	70	100	134	42
Natural sweeteners	80	100	135	42
Synthetic sweeteners	110	131	172.5	42
Preservatives	50	75	105	41
Antioxidants	62.5	92	115	41

tion could markedly reduce the incidence of food-poisoning outbreaks throughout the EEC. With the bulk of food-poisoning outbreaks being associated with consumption of fresh/chilled foods, the forecast doubling in the consumption of such products over the next 15 years makes the widespread adoption of irradiation techniques even more desirable. Clearly, if irradiation is going to be available in the year 2000, considerable effort is required by the industry to counter the ever-stronger voice of the consumerist lobby.

The impact of biotechnology specific to the food industry will not be as great as in the agricultural sector although, clearly, some developments in the agricultural sector will indirectly benefit the food industry. For example, the use of cell culture techniques and cloning will result in an improvement in the quality of a number of primary raw materials used by the food industry. Cyclical shortages of some food ingredients, notably natural gums, could also become a thing of the past through the exploitation of genetically modified plant strains that are less susceptible to climate change. Possibly the most important contribution of biotechnology to the food industry will come from its use to modify raw materials, e.g. fats, starches, giving them improved processing and sensorial characteristics. One could envisage the modification of fats by enzymes or microorganisms so that they are less harmful from a dietary standpoint.

Turning to traditional processing techniques, the growing importance of fresh/chilled foods in the market is supported with a doubling in consumption forecast by the year 2000. Over the same time period, consumption of canned foods was forecast to decline 30%, which is in line with the views of many research centers. For frozen foods an increase of 10% up to 1990 was predicted, but as to 2000 there was some uncertainty, as the further development could be determined by the likely future contribution of novel food-processing techniques. The average forecast was an increase of frozen food consumption of 25% up to 1995 and around 35% up to 2000, but the variation was from minus 3% to plus 88%.

As to the likelihood of scientific and technical innovations over the next 15 years, the inherent and not surprising conservatism of the food industry will tend to be a limiting factor. The growing consumer awareness and concern regarding food-processing practices, e.g. irradiation, could also limit innovations.

While the general trend is likely to be increasing levels of scientific and technical sophistication, there could well be a growing consumer demand for products manufactured by traditional methods. This itself would create the need for a more skilled, knowledgeable and versatile work force.

With the forecast increasing dominance of large and technically more sophisticated companies, consideration should be given to small operations, which do not enjoy the same economies of scale. To this end an increase in the number of regionally centralized facilities, offering the latest processing techniques and advice, should be encouraged.

At the production level, developments in the area of process control, and more widespread use of mathematical modeling techniques will enable the food manufacturer to produce a more uniform product in terms of taste, texture, color, etc.

New packaging materials and techniques will result in the appearance of more fresh, portion-controlled food products, catering not only for the consumer at large but specifically for the increasing number of elderly consumers. The prime aim will be to provide all consumers with increasing levels of convenience.

Table 4. Greater variety of food products

Worldwide exchange of foods
Worldwide exchange of men
Activities of the food industry
Packaging, storage, and distribution of foods of different processing degree
'Fresh': special packaging, cold storage, CA storage
Fabricated: specific processing, packaging, storage
Preserved: safety with reduced processing, packaging, storage
Main problems
PPP (product/process/packaging)
TTT (time/temperature tolerance)

Table 5. Increasing proportion of food products which are processed less intensely

Consumers prefer nontreated, natural-like foods
Consumers, partly because of ignorance, are critical of chemical and some physical preservation methods
Activities of the food industry
Product suitability, choice of varieties, state of ripeness, and product stability have to be adapted to the expected residence time in commercial facilities
Direct coordination among production, processing and trade
Monitoring of the temperature-time history according to HACCP
Main problems
To establish and maintain appropriate conditions with regard to packaging and storage, mainly for nonprocessed and chilled foods

The general impact of science therefore seems to be quite clear, but it should be useful for individual food manufacturers to recheck the situation as to their special activities from time to time.

Some Examples for Necessary Research and Development Activities

As already mentioned, some examples and their necessary technological realization can be detected. The motives or reasons, the necessary actions of the industry and the main technical problems are summarized in tables 4–9 [3].

Perspectives of Food Legislation: The Force for a Common Food-Industry Policy

What characterizes the completion of the internal market is a 'new approach' advocated by the Commission and accepted by the Council which

Table 6. Growing interest in stable foods of improved quality

Increasing demand for fabricated foods
For certain situations long shelf lives necessary
Activities of the food industry
Selection of raw materials/suitability of raw materials, product composition, and processing conditions adapted to quality requirements of the end products
Problems of suitability regarding classical preservation methods such as heat treatment, cold application, dehydration
Appropriate product composition especially of multicomponent products
Consideration of the PPP concept
Main problems
To optimize preservation steps as part of processing in consideration of the kinetics of changes of important ingredients and properties.

Table 7. Increasing convenience in the presence of high quality

Increasing demand for convenient foods
Preference given to convenient and attractive products
Activities of the food industry
Manufacture of easy-to-handle, ready-to-heat or ready-to-serve foods of high quality
Product concept allowing easy handling
Utilization of the advantage of multicomponent products
Use of gentle preservation steps to protect quality
Main problems
Variation of the product composition to achieve favorable rheological and thermal properties for an optimal processing

consists, in particular, of eliminating trade barriers by means of mutual recognition of national regulations, to which harmonization is added via community directives in vital areas such as health and consumer protection. This approach is supported by a large body of Court of Justice case laws, which gave article 30 its real scope by defining a basic rule, that any product legally manufactured and marketed in a member State must be accepted in all other member States. And it is only when this principle cannot be applied that community legislation will be necessary. Thus, in its 'White Paper bis' the Commission indicated that the Community legislation of foodstuffs must only include rules justified by protection of public health, the need to inform consumers and protect them in other areas besides health, fair trade, and the need to ensure official control.

Table 8. Development of products adapted to technical standards of household equipment

Preference is given to products which can be handled easily according to the technical possibilities in the household
Increasing demand for products which can allow creativity in the household kitchen
Activities of the food industry
Adaptation of the product properties to technical facilities in private households and to further treatments as desired
Useful and safe interruption of the processing chain
Adaptation of the product properties to technical facilities in private households and to further treatments as desired
Main problems
Industrial production should be aware of technical household facilities and indicate correct and workable instructions for further treatment

Table 9. Development of products of high physiological quality

Trend towards health foods
Trend towards foods without 'foreign' substances
Activities of the food industry
Development of products in compliance with recommendations of nutritionists
Physiological quality depends on selection of raw materials and their composition and combination
Supplementation and enrichment is necessary and useful in specific cases only
Control of recommended values
Main problems
Optimization of a product's quality potentials and preservation of physiologically important components

It is worthwhile that the European food industry has to keep a watchful eye on the development in the area of food legislation, to be able to support the individual companies through their national representation with the relevant information, and to present the views of the food industry in the relevant EEC institutions. Food legislation defines the frame in which the food industry can develop its activities. And, without going into detail, it is clear that the already very difficult area of food legislation with respect to the situation in the member States will become even more confusing, as there is no tendency in the Commission to extend the 'horizontal' by a 'vertical' harmonization in certain commodity areas. The Confederation of the European Food and Drink Industries (CIAA) therefore represents the views of the industry, as legislation forces to set up a common policy. Some of the present views of the CIAA should be mentioned to indicate a rough draft of this policy [4].

In order to reach the goal of free circulation of products, the Commission proposes, in its White Paper, a new strategy which would combine the principles of mutual recognition of national rules and standards, based on articles 30–36 of the EEC Treaty, with a new approach for the harmonization of legislations based on article 100A, which was introduced in the Treaty of the Single European Act.

The EEC food industry fully endorses this strategy but insists, with respect to so-called 'vertical' harmonization, that the Commission promises to examine, for certain products, any request made by a sector of industry if there is a consensus as to need for a Community vertical regulation. The same applies to a unanimous request by a sector for the abolition of an existing one.

With respect, in particular, to the four areas for which the Commission proposes to adopt directives on harmonization, CIAA's policy from the outset has been and always will be close cooperation with the Commission's services in drawing up drafts likely to achieve the aims set forth, but which also take into account the concerns of the food industry as well as market realities.

As to protection of public health, the food industry has always felt that no matter what problems it may be confronted with, health protection is the highest priority in respect of which all other considerations must take second place. It has therefore actively cooperated with the Community institutions and other bodies such as the WHO/FAO Codex Alimentarius which deal with this matter, in order to draw up the necessary rules to carry out this aim. CIAA has always supported, supports and will continue to support the Commission in its efforts to set up a high level of protective legislation on consumer health.

Currently, the Community is setting up such legislation, in particular for food additives, materials and objects intended for contact with foodstuffs, flavorings and irradiation.

(1) Food additives: Having noted that it is in this area that most barriers affecting intra-Community trade in foodstuffs are to be found, CIAA supports the Commission's efforts at setting common rules concerning the use of additives.

(2) Materials and objects intended for contact with foodstuffs: CIAA's position has always been that the responsibility should be shared among the

different sectors concerned (manufacturers of polymers, producers of packagings, and food industries), and that the global migration limit proposed by the Commission was not an acceptable solution. In an effort to compromise, however, CIAA is trying to formulate an answer which will satisfy the different industrial sectors concerned, and hopes that the Commission will take this into account in the text it is preparing on materials and objects in plastic.

(3) Irradiation: Faced with divergent national legislations, the food industry is pleased that the Commission has proposed a directive aimed at the authorization, under specific conditions, of the treatment of foods and food ingredients by ionizing radiation. Indeed, this scientific processing procedure is one whose total safety is recognized by scientists at both Community and world levels, and one in which important sectors of the food industry are greatly interested.

(4) Organic products: CIAA recognizes that in the EEC countries an increasing demand exists for products cultivated organically and that several food industry sectors are interested in the development of this market. CIAA is therefore, pleased, that the Commission intends to establish Community regulations to harmonize the conditions of production and marketing of these products and, in particular, to regulate the use of terms such as 'organic' and similar expressions. CIAA feels that only agricultural production, to the exclusion of processing and conservation operations, should be regulated by Community measures in this area.

As to consumer information and protection, the Commission states that there are two possible approaches. One would be to develop very detailed regulations on the composition and manufacturing characteristics of each food product ('recipe-law') and the other is based on the fundamental idea that, insofar as the buyer is validly informed of the nature and composition of products, it is not necessary to define these aspects by law, unless health protection makes it necessary. The Commission, followed in this by the Parliament and the Council, resolutely chose the second approach. The key element in the description of this second approach is adequate consumer information, hence the vital importance of labeling. This is, moreover, abundantly confirmed by recent Court of Justice case law.

With respect to public checks, the Commission presented to the Council a proposal for a directive concerning public checks of foodstuffs on which CIAA expressed a generally favorable opinion. Indeed, CIAA feels that approximation of national legislations in this area is an essential element of the new approach to food legislation followed by the Community, while the differences currently existing between national legislations in this area act as barriers to the free circulation of foodstuffs. Indeed, if the consumer is not convinced of the perfect safety of imported products, free exchange in the internal market will be an illusion. CIAA therefore insists that national procedures for carrying out checks be equitable so that conditions of consumer protection and fair trade are equivalent throughout the Community, thus allowing free circulation of food products. Moreover, cooperation must be developed between the national services carrying out these checks in order to facilitate exchange of information on national legislations and related procedures.

For a successful development as well in the year 2000, it is essential for all individual food companies to have the knowledge as to the existing food legislation, but it is well to consider the tendency of the future development. This is one part of the so-called quality assurance.

Quality Assurance: A General Strategy for Each Individual Food Company

The quality of our foods has reached a very high level today. Our foods are safer than they have ever been before, despite the general discussion on food quality, despite various 'pseudo scandals', and despite a few genuine scandals, which unfortunately can never be completely avoided. A certain residual risk will always have to be accepted. However, it is the task of the food industry to reduce this risk with the possibilities and means available today, and that is why quality assurance is more important than ever [5].

Quality is not a well-defined term. Simplified, quality is conformance to requirements. But what are these requirements? Again, they are not exactly definable in general valid changing attitudes. Progress in science and much improved analytical techniques allow more new information, sometimes without any possibility for a well-founded interpretation. The term quality is dynamic and therefore has to be separated into the most essential partial qualities or quality values, as shown in table 10 [6].

To realize a desired food quality in practice, not only the food product itself has to be taken into consideration. Furthermore quality of people, quality of systems, quality of services, quality of technology have to be envisioned. Quality is not the effort of an individual or a group, but a total commitment of an entire organization. It needs management.

Table 10. Food quality, partial qualities and quality values

Overall quality	Partial quality	Quality value
	–	Market value
	Nutrition-physiological quality	Nutritive value
		Health value
Food quality		
	Technological quality	Convenience value
	Sensorial quality	Taste value
	–	Ecological value

What is Quality Assurance?

Quality cannot be second to any other activity of a company. It has to be part of all the company activities. The quality of a food product is only as good as the strategy to be taken as a basis. Quality needs management, quality assurance translates the theoretical needs into practical strategy.

Only if top management understands the necessity and possibilities of quality assurance and takes it as an integral part of the strategy of the company will there be a chance to instal the quality assurance system in an effective manner. The scope can be seen as a hierarchy of 4 essential elements: (1) quality policy; (2) quality management; (3) quality control; (4) quality documentation.

As a profession, quality assurance must advance to the highest level of management according to the quality policy. As a function, quality assurance must be positioned to share equal recognition with other departments to achieve a stronger voice for quality assurance throughout the organization, a quality assurance status commensurate with the company quality goals, and a projection of quality as a vital element of the company philosophy.

The basis of quality assurance is a systematic concept, which possesses a general validity. Handbooks of quality assurance being such a general concept exist. But each company has to establish its own very specific quality assurance system, in order to realize the very important step of quality control at all relevant stages. Finally, documentation is the evidence, first for the company itself but also for the necessary external dialog with commissioners, customers, government agencies, official food control, trade organizations, consumers.

Elements of the Quality Assurance Concept in a Practical Situation

In practice, all aspects have to be considered as to a possible concern of quality. Therefore, to realize the general quality assurance concept it is necessary to look at the individual products from the planning or development stage until the distribution within the market. For all situations 4 elements can be established for a further detailed layout of the specific quality assurance system: (1) product idea/product planning/product development; (2) production and/or purchase of raw materials, food ingredients, packaging materials; (3) production and packaging of food products; (4) storage and distribution of food products.

In the planning and development stage of a product all guidelines relevant to the evaluation of foods have to be included, regardless of whether these are legal, other standards, guidelines, or recommendations for the evaluation of foods and materials coming into contact with them. This first theoretical approach must be based on the latest scientific knowledge, legal standards and own quality guidelines. It has to include all aspects relevant to the quality in the later production stage, as for instance: (1) setting up specifications for raw materials and components for the final product; (2) hazard analysis as to the potential production of the product; (3) determination of critical points which have to be controlled; (4) establishing of quality criteria and their physical, chemical, enzymatic, microbiological and sensorial assessment; (5) complete documentation of the practical development procedure; (6) identification of processing steps, storage conditions and distribution measures; (7) iterative establishing of the final development concept.

Only if, during this theoretical approach, all desired and necessary points are defined in a safe way and a catalog of basic data is assessed, can the practical product realization begin.

The first central area of industrial production comprises purchase of raw and packaging materials. Product related specifications are the basis of raw material delivery contracts. Of special relevance in this context are naturally occurring substances in foods of plant and animal origin, residues of agricultural chemicals, feed additives, veterinary drugs, environmental contaminants, measures during agricultural production. To meet the specific characteristics, it may be necessary to make special cultivation contracts, which include, for instance, selection of varieties, crop rotation, application of fertilizers, pesticide protection, harvest criteria, and post-harvest operations. Therefore, close cooperation and consulting are helpful in order to realize the planned quality of the necessary raw materials. The same, of course, is valid concerning packaging materials. Raw materials, ingredients and packaging materials must be controlled at the earliest possible stage and if they meet the specific quality requirements they may then enter the production process.

Production and packaging of food products under the aspect of quality assurance must consider products and materials, processing, and personnel/staff. As already stated, only the detailed and comprehensive knowledge of every single raw material and packaging material, as well as their potential changes in combination with a reliable and effective inspection guarantees the production of safe food products of the desired quality. To control the different unit operations of the whole process it is necessary to set up optimized process conditions, to carry out on-line control measurements where possible, and to observe the environment of the production lines. Finally, personnel involved in production and packaging have to be trained and adequately informed, as only a motivated person will understand the necessary activities and duties prescribed according to the specific quality assurance system.

Quality assurance as described here is a wide-ranging concept which identifies and is related to all those matters which individually or collectively influence the specific quality of a product. GMP is incorporated as well as other factors and has two complementary components, namely effective manufacturing operations, and effective exercise of quality control. GMP requires that every aspect of the manufacture is fully specified in advance and that all the resources and facilities specified are in fact provided in the right quantities, at the right times and places, and are utilized as intended. Effective food control requires well-qualified and appropriately experienced food control management with adequate staff and facilities to carry out all the relevant inspection, sampling and testing of materials, monitoring of process conditions and relevant aspects of the production environment, providing rapid feedback of information, thereby enabling prompt adjustment or corrective actions to be taken. GMP is a dynamic and adaptable concept, and an evolutionary mechanism by which overall improvements can be made and maintained.

Safety and Success through Quality Assurance

Quality assurance is not a new magic word. Most food companies already do something in the field of quality assurance. But it is necessary to consider a general valid, comprehensive approach with the different elements or bricks, which can be used for an individual quality assurance system.

Quality assurance means to produce on the safe side. One single product recall will cause loss in confidence, loss in image and loss in money.

Quality assurance is of fundamental importance to maintain or in some cases regain the confidence of the consumer in the products of the food industry.

Quality assurance is necessary to allow a successful production of high-quality products.

Quality assurance is the guarantee to adapt to the changing boundary conditions determining food quality.

Outlook

If all these points which have been discussed are kept in mind and supplemented, if necessary, the food industry should not only survive but enter successfully into the next century. But there is still one aspect to be considered.

Today there are many new products being introduced as well as a few innovations. However, many of the innovations are based on developments and patents granted 10–20 years ago. Most companies have cut back on research and development, sometimes dramatically. The food industry spends the lowest percentage of sales on research and development of any major industry, and the situation of even food science departments at universities and government research centers is characterized by reduced budgets. Besides that, research and development in industry is concentrated in a few companies.

All necessary actions of the industry described need research in different areas such as basic research and applied research in the narrow field of food science, as well as research activities in a more general sense in adjacent fields of interest. Perhaps not all the possibilities of information and cooperation have already been dealt with completely, but a more prepared attitude to support research and development will be necessary within the management of food companies. Although the food industry, because of its very different structure with many small and medium-sized companies, cannot be compared directly with other major industries, development has always shown the fundamental need of a well-defined and safe basis. This has to be realized today in order to harvest successfully tomorrow, which means sometimes 10 years or even more in the future.

References

1 IGLO-Forum: Die Deutschen und ihre Ernährung –Vorteile und Nachteile. Hamburg, 1989.
2 Young JN: The impact of food science and technology on the EEC food industry in the next 10–15 years. FAST activity: The future of the food system, 1987.
3 Paulus K: Can market and consumer trends as well as nutritional requirements be realised technologically? Food Congress 87, Helsinki. Kemia – Kemi 1987;14:1049.
4 Confederation of the Food and Drink of the EEC (CIAA): Elements of an EEC food industry policy. Bruxelles, 1989.
5 Paulus K: Maintenance of food quality: More important than ever. Food Tech 1989; 3:11–14.
6 Paulus K: Food quality assurance: An integrating system of the food industry. 14th Int Congr Nutrition, Seoul, 1989.

Prof. Dr. Klaus Paulus, Natec Institut für naturwissenschaftlich-technische Dienste GmbH, Behringstr. 154, D-2000 Hamburg 50 (FRG)

Lindroth S, Ryynänen SSI (eds): Food Technology in the Year 2000.
Bibl Nutr Dieta. Basel, Karger, 1990, No 47, pp 87–100

The Capability of the Food Industry in a Small Country to Utilize Technological Innovations

Juhani Olkku

Oy Lahden Polttimo Ab., Lahti, Finland

Food Systems

Food systems provide nutrition for human beings. Systems range from local collecting, fishing and hunting to specialized agrifood businesses with multinational production, processing and distribution chains. What can be considered to be an innovation in a food system depends on the degree of development and complexity of the system itself, on the society with its social system structures, and on the available technology and technical support system infrastructure. In other words, what constitutes a true innovation on a local scale in a developing country may not be considered to be an innovation in a developed country with effectively functioning infrastructures.

The food industry has evolved from home preparation of food through specialization and division of labour to a subsystem in the food system. The development has taken place mainly in this century, and according to Slater [1] it has four areas: (1) 1900–1925 evolution: emerging science of food preservation and safety evolves into factory processes; (2) 1925–1950 mechanization: batch processing converts into a mass production, continuous flow industry; (3) 1950–1975 diversification: new product and package developments spur widespread marketing and distribution; (4) 1975–2000 optimization: systems approach optimizes food production to meet human needs on a global basis.

The five human needs in food systems design are: (1) the need for palatable food; (2) the need for nutritious food; (3) the need for safe food; (4) the need for storable food; (5) the need for low-cost food.

As shown in figure 1, the food industry processing system is connected upstream to the raw material procurement system and downstream to

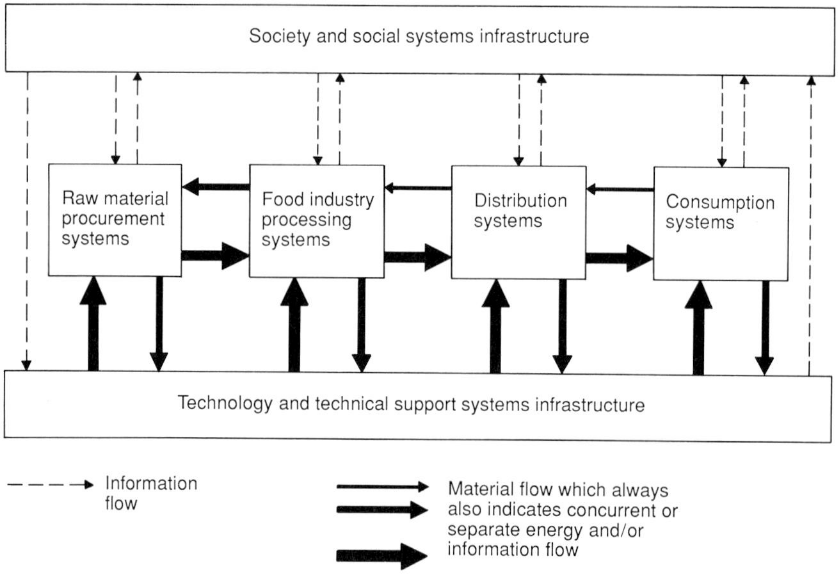

Fig. 1. Interdependence of food industry processing systems and other systems when supplying food for man.

distribution and consumption systems. The food industry processing system can be considered to be composed of two stages: (1) a primary stage, processing the products from the raw material procurement system to raw materials for (2) the secondary or completing stage, manufacturing the products for the distribution and consumption systems.

Slater [2–5] has suggested that the integrating factor in a food-processing system is the material handling and that interphases between system elements are critical for the proper functioning of the entire system.

On the horizontal level, raw material procurement systems precede the food industry. Agricultural systems are a dominating raw material source. Whereas in many a developing region there is a shortage of food, the overproduction of farm products in developed countries will force the utilization of cereal grains, oil seeds and legumes and/or the use of farmland and its products for nonfood applications. The latter trend will be enhanced by the application of biotechnology to plants, which also shows promise in the solution of food shortages in the long run. Target-oriented plant and animal breeding for various purposes with the help of new biotechnical systems will be the rule. Aquaculture will also develop to a more mature utilization stage.

In the future, food industries will exercise an ever-increasing control over their raw material development and procurement. Part of the products from the raw material procurement system are and will be returned to agricultural usage from the food-processing industries in the form of feeds and feed supplements whereby the utilization of the biotechnological techniques will also have an impact.

Distribution systems and consumption systems follow the food industry along the flow of food products. The distribution system can be open – i.e. freely competitive – or closed – i.e. governed by a few dominant distribution chains, as is the case in Finland. As these commercial chains forming a quasi-competitive closed system also operate in agribusiness and interindustry trade, they exercise an indirect control over the food industry. Consumption systems range from home preparation of food through restaurants to institutional feeding. The balance of various sectors changes with increasing affluence, available technology and the development and changes in society.

The infrastructure of social systems is determined by the immaterial factors emerging from the fact that each society has its own cultural and structural backgrounds. Eating and enjoying food is normally a social occasion in any culture.

Our conception of food emerges from the physical and chemical factors of food properties through a series of conceptual and simultaneous observations by our senses. Moreover, it is most strongly influenced by our mind and our models and concepts of proper food. These concepts are formed by social learning, i.e. what is considered to be good and appropriate, and by our own experimentation, feelings and sensations. All in all, food is a necessity which is immersed in deep and strong emotional factors.

Society and its systems form a framework within which an industry operates. Thus the structure and changes of these systems also affect the food industry, as the following developments and trends indicate.

(1) EC 1992, trading systems such as GATT, and the outcome of the present turmoil in Europe all present examples and prospects for widespread changes.

(2) Nutritional concepts are changing and developing in the light of new knowledge, sometimes having an irrational and inconsistent effect.

(3) The green philosophy in products we use and the environment we utilize will spread with the growing affluence.

(4) For want of predominant paradigms, societies will have segregated pressure groups which will come and go.

(5) Legislation – both national and international – is ever-increasing, with product liabilities and other stipulations coming into force.

(6) New national and international standards are created and will be applied extensively.

(7) We are moving from villages, nations and regions to true global concepts which in turn will enhance the influence of one's own cultural roots and the need to understand them.

Innovation

Innovation as a phenomenon, art and science has attracted increasing attention. Understanding the nature, basis and phases of innovation and the ability to control the innovation process will be a crucial success factor in the future.

Drucker [6] lists seven sources of innovation, divided into two groups:

(a) Sources internal to an industry or other system: the unexpected success, failure, outside event; the incongruity between reality as it actually is and reality as it is assumed to be or as it 'ought to be'; innovation based on process needs whereupon a process is broadly considered to be any system composed of consecutive actions; changes in the structure of industry or market that catch everyone unawares.

(b) Changes outside the enterprise or industry: demographics (population changes); changes in perception, mood and meaning; new knowledge, both scientific and nonscientific.

Drucker considers the last change, knowledge-based innovation, to be the hardest to understand and to bring about. It is, however, the topic of this paper, so we will continue to survey it.

Quoting Drucker [6], the basis of 'systematic innovation consists of purposeful and organized search for changes, and is the systematic analysis of the opportunities such changes might offer for economic or social innovations'. He stresses that innovation is application and dissemination, and it is a process which can be managed.

Roberts [7] starts his excellent treatise and review of innovation with the following words: 'Technological innovation can alter the competitive status of firms and nations but its purposeful management is complex, involving the effective integration of people, organizational processes and plans'.

Innovation is composed of two parts [7]: (1) the generation of an idea or invention, and (2) the conversion of that invention into a business or other useful application.

He continues by stating that 'using the generally accepted (broad) definition of innovation – all stages from the technical invention to final commercialization – the technical contribution does not have a dominant position.'

Roberts' short definition of innovation is: innovation = invention + exploitation. He classifies (technological) innovations into product, process and systems innovations.

Innovation can be seen as a multistage process, often with several iterative steps between phases. The process and some innovation illuminating factors condensed from Utterback's [8] ideas are described below.

According to him, each company has an innovation potential which is a function of its environmental factors, such as: (1) economic, social and political factors; (2) development stage of technology, and (3) information about technology. A company's own characteristic features such as resources, personnel, and communication and decision structures determine the possibilities of utilizing its innovation potential.

He divides the innovation process into three major stages: (1) idea generation; (2) problem solving and development, and (3) realization and diffusion.

Some observations about the idea generation:

It is a synthesis of different information which already exists including market or other needs and the technology needed to solve them.

60–80% of all innovations are based on market pull (which according to Walton et al. [9] will, in contrast to science-push R&D effort, require the parallel recommendation of patent strategy, a development strategy and a commercialization strategy during the programme definition in order to appropriately define and guide research).

Process innovations are generally adapted from other companies (i.e. there are only a few lead users of new process developments).

The cost of initiation and development of an innovation is 15–30% of its utilization cost. (A decade rule of expenditure derived from the US electric equipment industry states that 1 dollar spent on basic research needs 10 dollars for development work which in turn needs 100 dollars for plant construction and marketing of the products.)

The source from which the idea is born is frequently communication from outside the normal surroundings.

Some observations about problem solving and development:

This stage consists of setting specific technical goals and the development of alternative solutions in order to meet the goals and is iterative by its nature.

Information flow is very important at this stage. It has been found that information flow within a company is over 50% oral. Thus, a key role in the process is the gatekeeper, a person with good contacts and information.

Project groups and technology transfer units are often used at this stage. Such groups should include a project champion (or the engine pulling the train) and specialists of various needful fields.

In Utterback's [8] model, the completing stage in the development process of an innovation is realization and diffusion. Realization includes the development of production bases, process design, investment in production facilities and beginning of marketing, which are all needed in order to bring the new solution or idea to the market. Diffusion to the surrroundings begins with the presentation of the innovation and is a two- or multistage process. There are user gatekeepers or lead users [10] who are willing to adopt new products, services or processes. Thereafter, the dissemination or the development of market volume will follow the logistic 'S-shape' curve when other users adopt the innovation. Frequently, one has a tendency to be optimistic when considering the efforts, resources and time spans needed in the development and commercialization of a technological idea or invention. Yet one of the key factors of future success will be the ability to master and control the innovation process. The picture given above is simplified. More detailed presentations are readily available in the literature cited.

Roberts [7] sums up the innovation process in the following way:

The overall management of technological innovation includes the organization and direction of human and capital resources effectively toward: (1) creating new knowledge; (2) generating technical ideas aimed at new and enhanced products, manufacturing processes and services; (3) developing those ideas into working prototypes, and (4) transferring them into manufacturing, distribution and use.

Given that all essential technology is mature for application, it will take about 5–7 years to develop an idea or invention to the dissemination stage. If technologies to be developed need research and scientific work, a new factor, science waiting time, which can delay the process substantially, will enter into the already complicated system picture [11]. According to Drucker [6], it takes about 25 years to develop a scientific finding to a technology which is ready for commercial innovations based on it.

In general terms the type and frequency as well as costs of developing an innovation depend on the maturity of the technology and of the industry utilizing the innovation. Roberts [7] maintains that most technologies move through the following evolutionary stages: (1) An early stage dominated by frequent product innovations, (a stage where it is easy for a specialist to make quick product innovations with fairly small investments). (2) A transition stage characterized by increased process innovation and the emergence of a dominant product design (a stage where development teams enter the picture, the investment and lengthening lead times grow and the technology will start to be taken over by large firms or firms growing from small start-ups to the size supporting the infrastructure and investment needed). (3) A mature stage featuring a much lower rate and lesser degree of both product and process innovation (it is a big boy's game and the development of newer, superior technology may be occurring simultaneously elsewhere).

Roberts concludes that the innovation strategy and technological resource allocations of firms differ markedly depending upon the stage of their primary technology. This is one factor emphasizing the fact that top management commitment is essential to assure the success of broad-based programmes aimed at establishing the development of effective product and process innovations.

Food Industry and Innovation

The food industry often acquires its new technology together with machines and production systems. Thus, technology and technical systems infrastructure is essential to the food industry and so is the ability to understand the development of technology in the supporting systems. On the other hand, supporting systems, such as waste handling, may impose new restrictions on or possibilities for the food industry leading to an innovation potential to be developed.

Many branches of the food industry are mature; the frequency of innovations is low and innovations are typically small improvements in raw materials, products, processes and systems. On the other hand, new technologies can add innovation potential and technology pull with an increased rate of product and process innovations. Such new technologies may be based, for example, on the application of fast-developing science on the molecular level, causing a technology push for biotechnological applications in raw material, product and process fields, or on the applications of developing material sciences influencing the packaging industry.

One branch of the food industry which could serve as a case is the brewing industry. It is an industry with long traditions and high professional skills. The technology has been steadily developing; wort gravities have been

Table 1. Processing times of beer with presently used high-gravity brewing (HGB) technique and with immobilized genetically improved yeast

Operation	Processing time, days	
	present (HGB)	potential continuous brewing with improved immobilized yeast
Mashing	1	1
Main fermentation	7	1
Storage fermentation	14	0
Finishing operations	3	1
Total days to produce beer	25	3

increasing and processing has changed from fermentation in two consecutive vessels to a single tank system. The product range has been developed to include new high attenuation or 'dry' beers in addition to the traditional beers. There have been improvements in the processing equipment and automation has been adapted, yet the technology is essentially unchanged.

Continuous methods of production are usually more efficient from the engineering point of view than batch processes such as production of wort and its fermentation. In the 1950s and 1960s continuous brewing processes were developed and used in England, Spain and New Zealand, for instance. The process technology then employed did not produce beer competitive in quality and all the breweries then utilizing continuous methods have converted or are converting back to batch processing. Since then, there have been several attempts to develop a continuous method, but no breakthrough has occurred.

At the beginning of the 1980s, microbial and enzyme systems based on immobilization were developed. At the same time, the present biotechnology boom began due to the development of microbial genetic engineering. The technology pull was there to reactivate the process development. Potential process time and cost benefits can be seen in table 1.

The development speed has been as follows (table 2):

In the early stages with small flow volumes, the brewer's yeast was immobilized in alginate beads which are soft and form a material transfer barrier for most yeasts. They are thus difficult to apply to large-scale

Table 2. Development speed of beer fermentation with immobilized yeast

Storage fermentation with immobilized conventional yeast			Main fermentation with improved immobilized yeast		
year	scale	litres/day	year	scale	litres/day
1985	laboratory	2.4	1988	laboratory	1
1987	pilot	240	1989	laboratory	5
1988	pilot	5,000			
Threshold approaches industrial large-scale application.					

Table 3. Relative volume effectiveness of beer fermentation methods

	Volume effectiveness
Present HGB	1
Immobilized yeast, main fermentation	5–6
Immobilized yeast, storage fermentation	6–7
Immobilized yeast, main and storage fermentation	8
Immobilized improved yeast, main fermentation (1 day fermentation)	10

processing in such systems. Adaptation of new carrier material, a solid adsorbent material readily available for other purposes since 1987, was the technical breakthrough needed for large-scale development [12, 13]. Such a combination of existing technologies is a typical case of innovation in development making progress possible.

The possible volume effectiveness to be gained with the immobilized yeast in various processes over the present HGB methods is shown in table 3. The last case implies that one would use a yeast improved by genetic engineering methods [14].

In the case above, the development of a possible process innovation is discussed. The new technology is approaching the industrial application threshold. A technology promising substantially improved processing benefits is ready to be transferred to industrial users. What would be the possible barriers to its applications? The brewing industry is very competitive and conscious of quality. Even if it can be shown that the beer quality will match the one obtained with the present methods, the technical benefits, if not displayed as a cheaper price to the customer, will not be an incentive to change the process. In the countries where brewing is established, the application lag time might be long. However, the growth of the industry today is in countries where, due to the early development stage, such an innovation may well be adopted. In the case of the genetically improved yeast, there is still the question of applying genetic engineering and the acceptability of such a technology by the end users, the general public. The brewing industry would not apply any technology which is not considered to be entirely safe and accepted. This will delay the application until the time is ripe.

Some General Trends in the Development of the Food Industry

An UNCTAD report [15] states: 'In the last ten years the changing economic atmosphere, rising costs and uncertain consumer acceptance of innovations have changed the product development of food firms towards new uses of agricultural products by fractionation, by applying biotechnical processes and by optimization of the use of by-products'. This trend will continue to be valid.

Some technology pull/push trends also affecting the development of the food industry are:

(1) Development of microelectronics including its application to automation systems and process instrumentation. For example, batch processing can easily be automated, and this pushes the batch-continuous limits to larger capacities than before.

(2) Biotechnology will affect raw material, processing and products at an increasing rate.

(3) Development in analytics will allow the deeper understanding of phenomena during processing. On the other hand, trace analysis of harmful substances will lead to complications as the true meaning of extremely small amounts of various substances is not understood. 'Pushing the panic button' phenomena will be with us for some time to come, and we'll have to learn to handle it.

(4) Information systems are developing rapidly and will affect various aspects of life.

(5) Energy will become a key limiting factor if all nations use as much energy as the developed world uses today.

(6) Total utilization of raw material will be needed.

(7) Systems science applications will be an answer to master the ever more complicated world.

The industrial structure will change to comply with the emerging trade patterns. Firms will operate on local, regional and global levels.

Technological Innovations and a Small Country

Most technological innovations in the food industry are based on technology transfer from the technology and technical support system infrastructure. The prerequisite for innovations or utilizing innovations made elsewhere is an existing and developing knowledge base of the biological sciences, food technology and food process engineering.

A key factor for a small country is a high standard of education at all levels. This covers schools and professional schools, intermediate level institutions, universities and the continuing education of people already employed by research institutes and firms.

High-level research is another key factor. It is done in universities and institutions, as well as in firms, and needs a properly functioning financing system to support it. In a small country the number of people available will limit the quantity of research that can be done. On the other hand, it is easy to locate the expertise and to utilize it quickly. National and international programmes and networking with firms and research bodies are essential. This, however, requires personnel and financial resources. Thus, clean-cut strategic thinking and technology utilization policy is most important.

Innovative technology pull/push from the supporting systems such as agriculture, packaging, transportation, machine and process engineering, information and distribution and food preparation systems will enhance the probability of innovations. The food industry acquires much of its new technology with its machines and processes. In a small country like Finland, where machine systems production is limited, the focus is on high application know-how.

The capability of the food industry in a small country to utilize technological innovations depends on the stage of development and on the tradition of the industry and supporting systems reflected by the richness and utilization ability of the knowledge base. A small population doesn't warrant

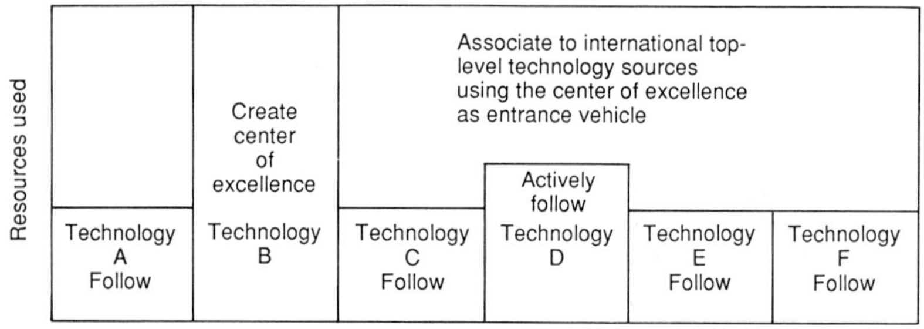

Fig. 2. Technology development strategy for firms or countries with limited resources.

major innovations in products and processes within the country. They need international dissemination scales and sometimes larger knowledge bases than are available in a small country. Thus, creation of knowledge-based innovations implies a strategy where the development energy is directed at a few selected targets where a critical mass can be concentrated in order to create a centre of excellence in the area. In other areas the strategy should be based on developing and maintaining a high enough level of technological knowledge and activity base so that a rapid and effective technology transfer from the sources of innovation can be realized. This model is illustrated in figure 2.

The status *follow* means literature and other information surveying and possibly a very limited research effort. The status *actively follow* means active research in the area aiming to raise the local level ready for technology transfer from the source of technology. The status *create centre of excellence* means a full effort to create a key technology superior to or at lest level with other technology sources, a technological competitive edge.

Today, information about technology is global and easily accessible, and it may be used to enhance one's own research efforts [9]. On the other hand, the odds are that the same ideas are worked on simultaneously by an increasing number of teams worldwide. Thus, the number of innovations is increasing even if the development time is shortened relatively little.

Perrino and Tipping [16] present the above facts in the following way:

'Competition for emerging markets and technology leadership is nothing new. What is new, however, is the intensity and globalization of that competition. Also, the dynamics of technology are changing:

Pace of technology is accelerating, raising the stakes and penalties for managing innovation, and requiring early warning and shorter response time to capture opportunities.

Specialization and systems requirements are both increasing, driving a growing need for interfacing people and disciplines, and integrating critical skills from wherever available.

Newer technologies are rapidly becoming pervasive, redefining competitive added value across a broad range of traditional markets and spawning major emerging growth markets.'

How can a small country survive and be competitive in the present accelerating technology-based innovation game? It must enhance its capability by deepening its basic knowledge base, by creating effectively functioning information utilization systems, by rational innovation policy planning and resource allocation, and by applying systems science principles in the field.

References

1 Slater LE: A global view of food process engineering; in Linko P, et al (eds): Food Process Engineering. Food Processing System. England, Applied Science Publishers Ltd, 1980, vol 1, pp 17–25.

2 Slater LE: Food plant automation – revisited. Food Eng 1975;47:37–55.

3 Slater LE: Programmable controllers: The missing link is here. Food Eng 1975;47:67–69.

4 Slater LE: The hierarchy of world food automation. Food Eng Int 1977;2:37–49.

5 Slater LE: FEI special report. Food materials handling: A systems integrating force. Food Eng Int 1978;3:37–41.

6 Drucker PF: Innovation and Entrepreneurship, Practice and Principles. London, Heinemann, 1985.

7 Roberts EB: Managing invention and innovation. Res Technol Manag 1988;31:11–29.

8 Utterback JM: Innovation in industry and the diffusion of technology. Science 1974;183:620–626.

9 Walton KR, Dismukes JP, Browning JE: An information specialist joins the R&D team. Res Technol Manag 1989;32:32–36.

10 von Hippel E: New product ideas from 'lead users'. Res Technol Manag 1989;32:24–27.

11 van den Muyzenberg L: R&D delivers products faster; in Johnson and Associates, SA (eds): The International Management Development Review 1989. London, Sterling, 1989, vol 5, pp 335–340.

12 Grönqvist A, Pajunen E, Ranta B: Secondary fermentation with immobilized yeast – industrial scale. European Brewery Convention. Proc 22nd Congr, Zurich. London, Oxford University Press, 1989, pp 339–346.

13 Kronlöf J, Härkönen T, Hartwall P, et al: Main fermentation with immobilized yeast. European Brewery Convention. Proc 22nd Congr, Zurich. London, Oxford University Press, 1989, pp 355–362.
14 Suihko M-L, Penttilä M, Sone H, et al: Pilot-brewing with α-acetolactate decarboxylase active yeast. European Brewery Convention. Proc 22nd Congr, Zurich. London, Oxford University Press, 1989, pp 483–490.
15 UNCTAD: The food processing sector in developing countries: Some recent trends in the transfer and development of technology. New York, United Nations, 1980.
16 Perrino AC, Tipping JW: Global management of technology. Res Technol Manag 1989;32:12–19.

Juhani Olkku, PhD, Research and Development, Oy Lahden Polttimo Ab., PO Box 14, SF-15211 Lahti (Finland)

Lindroth S, Ryynänen SSI (eds): Food Technology in the Year 2000.
Bibl Nutr Dieta. Basel, Karger, 1990, No 47, pp 101–114

Technological Development in the Food Industry in the Outgoing 20th Century

Walter E.L. Spiess

Federal Research Center for Nutrition, Karlsruhe, FRG

Utopias have been found much more realizable than was assumed previously. Today we are confronted with a question which is frightening in quite another way: how can we prevent Utopias from becoming reality?
Nicolei Berdjajew

Introduction

In what way will our social and technical world develop at the end of the 20th and the beginning of the 21st century – this has been a fascinating question. One interesting aspect among many others is the future development of nutrition of people in the highly industrialized countries. What will the food industry contribute to this development, and what is it expected to contribute?

Before the question of future development in the food industry is discussed, a few sentences should be devoted to the reasons for the present vivid interest in our future world.

Expectations and curiosity are nourished primarily from some spectacular changes which have been taking place for years in some sectors of our industrial spectrum, such as changes in the communication and information techniques, biotechniques and space techniques, to mention only a few. Observers furthermore expect a worldwide rise in technical innovations once the large European market is established. Not to forget that special revolutionary energy is widely believed to be associated with the turn of a millenium – to many a fact of nearly mythical nature. Programs like 'Aliment

2000' of the French government, or 'Men and Nutrition 2000', a German prognostic study, demonstrate the importance that is attached to the date.

What are the developments which may be expected in the food industry at this time of trend-breaking changes, great technical potentials and aggressive technical intelligence?

Considering the question matter-of-factly, i.e. replacing the spectacular date of turn of the millenium by a prediction period of 10–20 years, and re-orientating present ideas which have been mainly focusing on progress and revolutionary change, towards sensory and physiological requirements of consumers, one will soon recognize that – roughly speaking – there will be no profound changes until the year 2000 or, more sensibly, until 2001.

In the fields of consumer habits, social environment, production, measuring and controlling techniques, and storage and distribution techniques, however, several trends become obvious which may lead to an evolution of food production at least in some essential sectors.

Possible Influences on the Development of the Food Industry

Factors influencing the food industry in the years to come may be: social changes; changes in the ecological boundary conditions; new scientific knowledge of optimal forms of nutrition; changes in raw materials; technical innovations in the food industry, and innovations in related industrial fields and subcontracting industry.

These factors which may, in one way or another, contribute to changes in the food industry, will be exemplified below; factors relating to sociological changes will be discussed in the light of the present situation [1] in the Federal Republic of Germany.

Social Changes in Food Consumption and Supply

Demographic developments: Important to the quantity of food produced is the size of the population to be supplied, while age structure, size of households, educational level and kind of job largely determine the product spectrum. Available incomes are another influencing factor.

Population statistics of all European countries except for Greece show a stagnation, or slight decrease in population numbers (fig. 1). This implies an overproportionate increase in the number of old people (fig. 2) and will

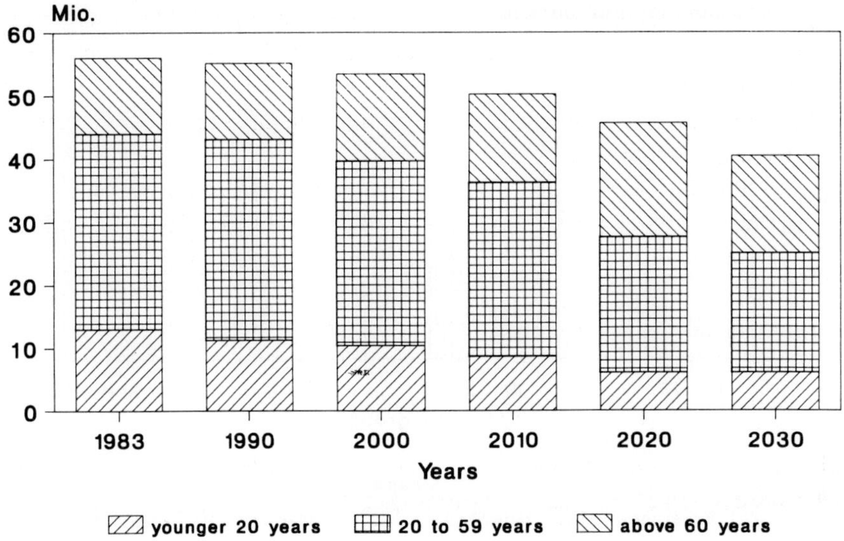

Fig. 1. Age structure of the German population, 1983–2030. Source: Statistical Yearbook 1985.

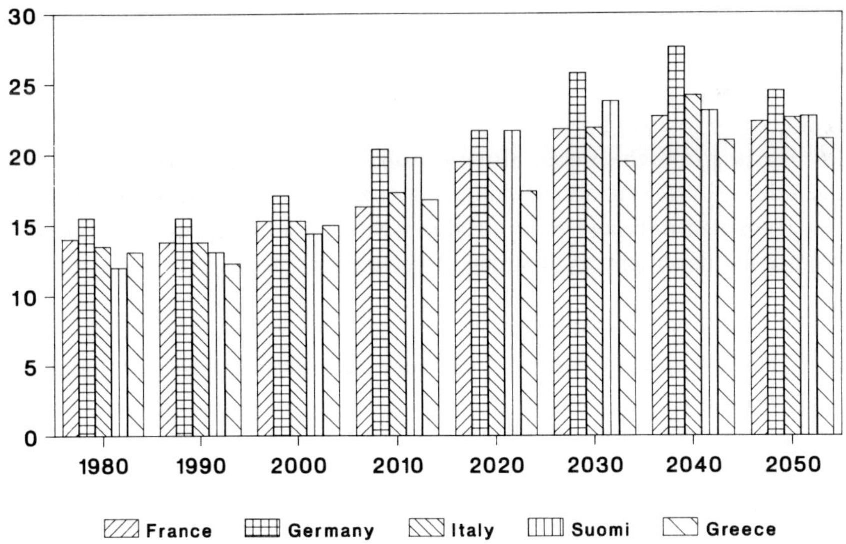

Fig. 2. Percentage of population aged 65 and over, 1980–2050. Source: OECD Index 1980.

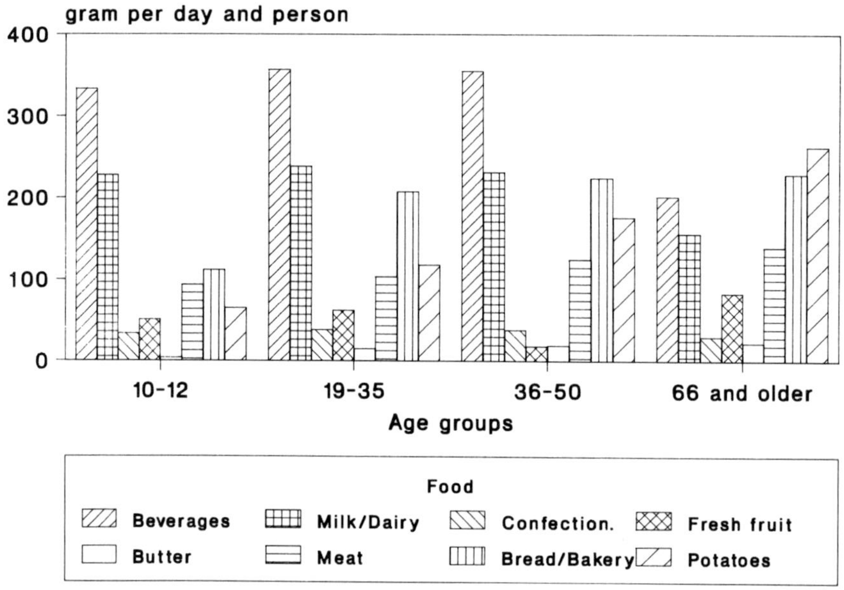

Fig. 3. Food consumption of men. Source: Nutrition Report of the DGE.

certainly lead to a restructuring of food supply in the sense of an adaptation to the preferences of old people which are different from the preferences of middle-aged or young people.

Starting from the present situation, one may expect a relative increase in the sales of meat, potatoes, fruit and vegetables in 2001, while sales of beverages and milk and dairy products will probably decrease (fig. 3). To predict the development of supply precisely according to the age structure presents some difficulties, because individual consumption habits depend not only on nutritional requirements, but also on growing insight into adequate nutritional behavior, adolescent conditioning, fashion trends and the influence of promotion. Household structure, education/job, and income determine the convenience degree of the food in the basket.

Due to a change in value judgments, life concepts and the way of living, the number of small households will continue to grow (fig. 4). Small households, so experience has shown, spend more money on high-quality convenience food (fig. 5). This suggests a continuing trend towards more convenience food and – what is even more important – towards a broader spectrum of convenience products. This trend will probably also be sup-

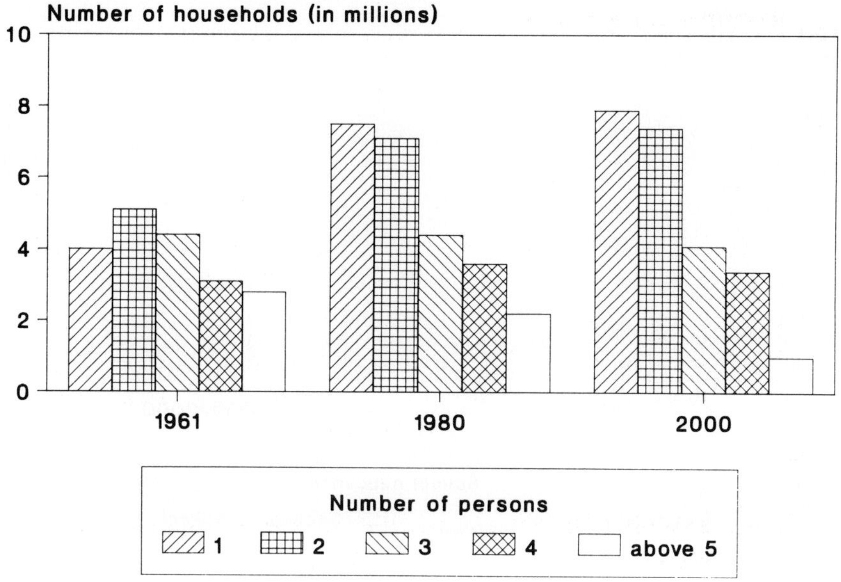

Fig. 4. Development of household sizes in the Federal Republic of Germany. Source: Federal Statistical Office.

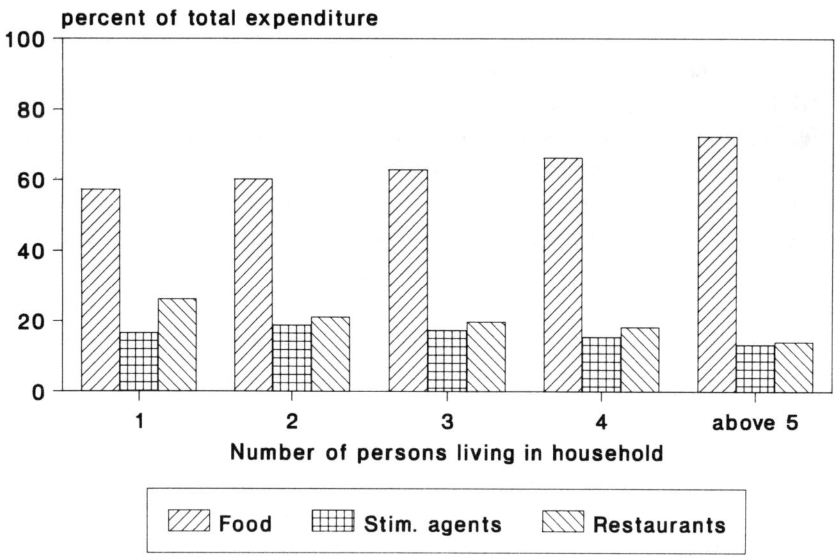

Fig. 5. Food purchases as a function of household size. Source: Statistical Annual Report on Nutrition, Agriculture and Forestry.

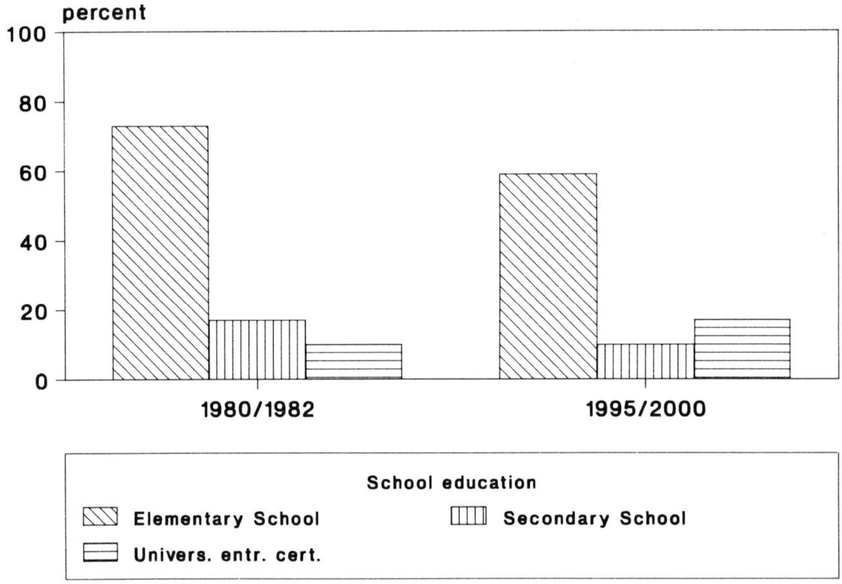

Fig. 6. Education level of the German population. Source: Federal Statistical Office.

ported by shifts in the educational levels (fig. 6) which are associated with changes in the professional structure, and by the higher available incomes (fig. 7).

Despite the recognizable shift from physical to intellectual work and despite growing awareness of health, there is no reduction in the energy intake by consumers (fig. 8). This fact which indeed gives reason for concern, can probably not be changed by external influences, since consumers, now as before, spend a relatively high proportion of their incomes on food and attach great importance to the sensory quality of the products – with the present nature (composition) of food, high energy intake therefore seems unavoidable.

As a result of educational measures, there will, hopefully, be a change in the kind of basic food consumed in the years to come. Trends, from animal protein to vegetable protein, from simple fat to vegetable fat of higher quality with less cholesterol and more polyunsaturated fatty acids, and from 'easily' digested products to products containing more dietary fiber, are supposed to increase.

Fig. 7. Development of the number of workers and employees in the Federal Republic of Germany. Source: Federal Statistical Office.

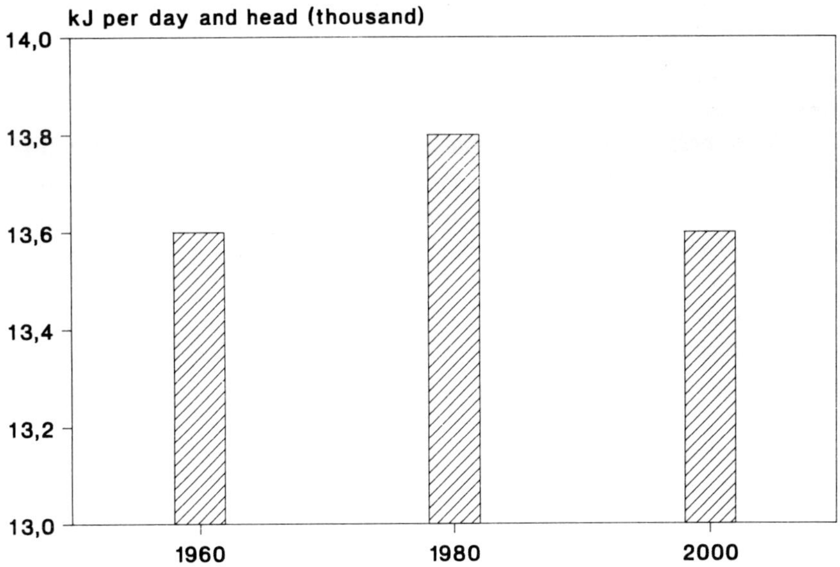

Fig. 8. Energy intake through food in the Federal Republic of Germany. Source: Federal Statistical Office.

Table 1. Purchasing behavior which is aware of the environment (about 45% of housewives)

Product	More seldom	More frequently
Sanitary detergents	××	
Air improvement agents	××	
Household detergents	×	
Bath foam	×	
'Muesli'		×××
Cornflakes/oat flakes		××
Cheese		×
Deep-frozen food		×

As far as available, highly caloric ingredients will be replaced by those showing lower energy values; prerequisite for the use of substituents with a low energy value such as sugar (saccharin, aspartam, etc.) and fat substituents (dextrans) is, however, that the sensory quality of products containing such substituents comes very close to that of the original products. In this case, products with a reduced energy value will probably gain considerable market shares.

Another sociological aspect influencing the commercial food supply is the awareness of the environment and the attitude of consumers towards the food industry and industrially manufactured food.

Consumers being aware of the environment (about 50% of all housewives regard themselves as being aware of environmental problems) prefer products with a high natural food appeal ('muesli', cornflakes, oat flakes, cheese) but also products whose external appearance was little changed (e.g. deep-frozen food) (table 1).

It should be mentioned in this connection that among German consumers, despite this high esteem of environment and nature, there is no general aversion against industrial products. Practically all age groups who decide what products to buy, accept industrial food in general; only a small group of consumers (1–2%) is completely discontented with industrial food. That about 90% of consumers are largely or fully content is lastly due to the very good image of industrial food. German consumers rely on the high quality and safety of trademark food products because they are subject to strict controls in which consumers have complete confidence (90% of consumers' opinion).

Nonspecific ranking of values and behaviors, respectively, influencing food intake and hence the industrial product spectrum may be outlined as follows:

The age group of Central Europeans between 40 and 50 who will be socially decisive in the year 2000 have never been starving and were brought up to be consumers. Together with certain sociodemographic facts such as increase in the number of small households, more leisure time, flexible working hours, but also longer journeys to work and a general reorientation of values, characterized by individualization, orientation towards the immediate presence, active ways of living (sports), and active media consumption, this may lead to a disappearance of the traditional eating culture. It should be underlined, however, that these conditions may, but need not necessarily be, stringent in this respect.

These restructurings may – or may not – favor fast-food menus taken at home or outside or may lead to food whose composition and appearance are subject to strong fashion trends. Fast-food components and convenience food, however, will no doubt survive.

What can certainly be expected is a refinement of the traditional ways of food intake – both of indigenous and adapted forms. Eating as a cultural activity will characterize the lifestyle of humans also in the future. This refinement of food intake will not be independent of industrial products; industry will have to supply the ingredients of the country-style and candlelight dinners, and also provide smorgasbröd and frutti di mare.

Influence of Nutrition Sciences, Ecology, Processes Showing Dynamics of Their Own, and of Transfer Processes on the Future Development of the Food Industry

Besides the aforementioned sociological factors, several other factors will influence food engineering and food. These factors will be outlined briefly in the following:

Nutritionists in general demand that three criteria be taken into account in any future nutrition programs. They are: (1) reduced fat intake, and especially reduction of cholesterol intake; (2) increased dietary fiber content in food, and (3) reduced consumption of animal and increased consumption of vegetable protein.

These demands, as mentioned before, have already been adopted by consumers. Possibilities of realizing them exist to some extent. The success of substitutes, as explained before, depends on how these components will be able to meet the sensory requirements of consumers. This is a challenge to scientists to utilize all possibilities offered by plant breeding and biotechnology, but also to food chemists and physicists.

Ecology: Future processing techniques must be orientated towards economic use of natural resources, i.e. optimal use and avoidance of environmental pollution. These strategies make it necessary that raw material obtained from agriculture is selected to fit this purpose. Some fundamental demands which have already largely been taken into account, are: (1) adequate shelflife of the raw material; (2) the material must be easy to clean and easy to peel (smooth surfaces); (3) low contents (no traces) of pesticides and environmental contaminants; (4) sufficient biological degradability of nonusable waste (no accumulation of toxic compounds in the waste).

For all these measures integrated planning is necessary especially with regard to possible consequences for and feedbacks to the environment.

Another problem to mention is the ozone problem which may influence the production, distribution and storage of chilled food. It must be considered as a severe challenge also to the food industry.

Some Selected Examples of Advanced Food Engineering Processes

The majority of innovations in the technical field will be due to the consistent further development of available processing techniques. This applies to the essential sector of thermal processing of food, but also to separating, concentration, cleaning processes at the beginning of, and packaging at the end of the engineering process.

For both ecologic and economic reasons whose interdependence will no doubt grow considerably, it can be expected that the use of water and energy in engineering processes is altogether reduced. Use of fresh water for cleaning processes will probably have to be minimized. Conceivable substitutes may be based on systems featuring air-water mixtures or water-sand mixtures in agitated air (air-water jet; water-sand jet). Efforts to clean waste water will gain in importance.

Ecologic-economic reasons will be decisive also for the use of direct heating processes such as microwaves for blanching. In vegetable processing, the gift of prophecy is not necessary to predict an increase in the use of *aseptic processing*. Developments will probably concentrate in three partial fields. Noticeable also for consumers will be the growing particle size of high-temperature/short-time treated products. Size lengths of particles at present are 10 mm at the maximum; in about 10 years, particle size lengths of 25–30 mm will present no problems due to advances in conventional heating systems or systems already in use such as ohmic or microwave [4].

With sterilization/pasteurization of liquids, improvements in quality preservation may result from processes in which concentrated material is heated by vapor injection and cooled by cold-water injection. These processes in which heating rates of 250 K/s and cooling rates of 2,500 K/s are obtained allow more precise temperature control than conventional HTST treatments [5].

To allow the utilization of possibilities provided by the aforementioned heating method and by ohmic and microwave heating, it is necessary, however, to further advance our understanding of structure and effect of enzymes, and to make full use of electronic possibilities of process control and regulation [6].

In the field of separation techniques it can be expected that in 10–15 years mechanical separation methods will have been advanced as far as to allow for separation of product components such as starch and protein without wet processing, and to obtain a high purity degree of individual components, depending on the mechanical decomposition of the product. Similar efforts are made to decompose fat- and oil-containing vegetable tissue, and to largely release the fat/oil phases without the use of solvents.

As far as the treatment of liquids (solutions, suspensions, emulsions, etc.) is concerned, the use of *membrane* techniques will play an essential role. It can be expected that microfiltration will become the method of choice to separate microorganisms, filtrate and clarify juices; ultrafiltration will be used to obtain protein fractions, amino acid components, enzymes, etc.; reverse osmosis to concentrate aqueous solutions. Present research suggests that membranes will altogether become thermally and mechanically more resistant (properties already existing in ceramic membranes), and that fouling problems which at present are a process-determining factor, will be overcome. Present attempts to find solutions to the fouling problem by reduced adsorption of membrane surfaces or enzymatic degradation of the cover layer seem promising. Improvement of the molecular weight cut-off (MWCO) is another interesting approach [7].

The possibilities of using electrodialysis and pervaporation plants will grow. For the latter, membranes of improved selectivity are being developed.

This will help to economize the concentration of alcohols, aroma compounds and the like.

Other processes which include a minimum of thermal stress, and which will, by the turn of the millenium, possibly be applied on an industrial scale, are extraction with supercritical CO_2, for instance to fractionize fat; furthermore, drying with heated CO_2 (associated with aroma recovery), or cold sterilization using ozone [8].

From among the large number of innovations to be expected, I wish to mention products resulting from restructured suspensions or emulsions as, for instance, soy protein products. The structurings are based on a precise analysis of naturally structured products; structure elements are accordingly combined and built into a network system, as in the brick and mortar approach. Structuring methods are to be found in classical processes like baking and freezing/thawing, and in new ones like extrusion cooking or spinning techniques [9].

As the last interesting field of potential innovations on the basis of available processes, the prospects of packaging will be mentioned briefly.

Novel packaging systems for chilled food, whose proportion in the total food spectrum will have increased considerably by the turn of the millenium, for fruit and vegetable products and aseptic products can be expected. Present development trends suggest that packages will be altogether less expensive; packaging materials will, to a much greater extent, be biologically degradable, or be edible.

Biologically degradable package materials – e.g. plastics with starch-containing components – may be used for dry products. Edible packages, e.g. on a casein basis, are suitable as covers over peeled/precut fruit and vegetable products. Such covers reduce moisture loss and inhibit browning reactions. Another kind of package for fresh produce shows scavenger properties and allows certain microclimates to be produced in a package [10].

Influences on the Development of the Food Industry by Know-How Transfer from Related Scientific Fields

In the last decades, food techniques received essential impulses from related technical and scientific fields, primarily from chemical engineering and bioengineering techniques. Food engineering has also profited from advances in the engineering industry and electrotechniques and electronics, respectively.

Advances in biotechnology first helped to lead traditional biotechnical processes such as sauerkraut production and brewing of beer out of stagnation and initiated a revolution in these fields. Furthermore, microorganisms and enzymes will be widely used to produce food ingredients such as aroma compounds and coloring agents, and to reduce undesirable substances such as nitrate, oxalic acid, etc. [11].

The use of enzymes has several advantages: (1) enzymes are already active at low temperatures; (2) reactions in selected substances caused by enzymes may be highly specific, and (3) food is only exposed to little stress, since enzymes are not used in extreme ranges of temperature and pH.

Immobilized enzymes prevent foreign proteins from remaining in the food. Since enzyme immobilization is application and process specific, new carrier-enzyme systems are required for each individual case. Usually, several immobilization techniques are used in combination.

The advantages are: (1) continuous processing; (2) reduction of processing steps to inactivate or separate enzymes, and (3) possible use of multienzyme complexes.

Enzymes, furthermore, are used in measuring techniques and for food analysis; however, enzymatic measuring techniques for process analysis are still relatively expensive and time consuming. Recent research has therefore been concentrating on the development of biosensors, e.g. to regulate fermentation or food-engineering processes.

Examples of biosensor use are: (1) catalase sensors for the determination of H_2O_2; (2) glucose oxidase for glucose measurements, and (3) monoamino oxidase to determine monoamines in meat.

Finally, two fields will be mentioned briefly which will certainly contribute to the scene in 10–20 years. One is the use of intelligent systems for the production of mass-produced articles with simultaneous product and quality control (computer-integrated manufacturing, CIM), the other concerns the use of robots. Especially with advances in image-analysis systems it will be possible to also have difficult work in the field of food processing, as for example, the cutting of carcasses, done by machines [12].

Independent of the kind of technical innovation, a potential of processing possibilities will be available by the turn of the millenium which will allow better and healthier food to be produced at a reasonable price. Whether this potential will be used or not, it will not be decided in 2000 or 2001, but in 1990. Today we must decide where to go and what risks we are ready to take in 10 years.

References

1 Mensch und Ernährung 2000: Gfk-Marktforschung GmbH und Co. KG, 1986.
2 Statistische Jahrbücher der Bundesrepublik Deutschland: Jahrgänge 1960–1980. Wiesbaden, Statistisches Bundesamt.
3 Statistisches Jahrbuch über Ernährung, Landwirtschaft und Forsten der Bundesrepublik Deutschland: Jahrgänge 1960–1988. Bonn, Bundesministerum für Ernährung, Landwirtschaft und Forsten.
4 Sastry SK: A model for continuous sterilization of particulate suspensions by ohmic heating; in Spiess WEL, Schubert H (eds): Engineering and food, vol 2, Preservation Processes and Related Techniques, p 270, Elsevier Applied Science, 1990.
5 Müller H: Direct sterilization/pasteurization (DS/P); in Spiess WEL, Schubert H (eds): Engineering and food, vol 2, Preservation Processes and Related Techniques, p 273, Elsevier Applied Science, 1990.
6 Kessler HG: Aseptic processing. Abstr 5th ICEF, Cologne, May 28–June 3, 1989.
7 Hallström B, Trägardh G, Nilsson JL: Membrane technology in the food industry; in Spiess WEL, Schubert H (eds): Engineering and food, vol 3, Advanced Processes, p 194, Elsevier Applied Science, 1990.
8 Budde A, Knorr D: Reduction of cholesterol in egg powder and whole eggs by extraction with supercritical carbon dioxide; in Spiess WEL, Schubert H (eds): Engineering and food, vol 3, Advanced Processes, p 159, Elsevier Applied Science, 1990.
9 Lillford PJ, Davies AP, Wilding P, et al: Manipulation of carbohydrate and tissue structures by process formulation; in Spiess WEL, Schubert H (eds): Engineering and food, vol 1, Physical Properties and Process Control, p 275, Elsevier Applied Science, 1990.
10 Labuza TP: 'Active Food Packaging' technologies; in Spiess WEL, Schubert H (eds): Engineering and food, vol 2, Preservation Processes and Related Techniques, p 304, Elsevier Applied Science, 1990.
11 Kiesser T, Bauer W: Use of enzymes in food production; in Spiess WEL, Schubert H (eds): Engineering and food, vol 3, Advanced Processes, p 437, Elsevier Applied Science, 1990.
12 Clark JP, Balsman WF: Computer integrated manufacturing in the food industry; in Spiess WEL, Schubert H (eds): Engineering and food, vol 1, Physical Properties and Process Control, p 781, Elsevier Applied Science, 1990.

Prof. Walter E.L. Spiess, Bundesforschungsanstalt für Ernährung,
Engesserstrasse 20, D–7500 Karlsruhe 1 (FRG)

Lindroth S, Ryynänen SSI (eds): Food Technology in the Year 2000.
Bibl Nutr Dieta. Basel, Karger, 1990, No 47, pp 115–117

Summary Remarks

Seppo Lindroth

Food Research Program, Ministry of Agriculture and Forestry,
Helsinki, Finland

The interest and active involvement of Finnish research and researchers, as well as research administration in European and other international cooperative work was described by Dr. Kuusi.

The impact of increasing consumer demand, internationalization of food production and trade, shortened life cycles of products and leaps in technological development have, in many countries, led to increased industrial and national interest in utilizing the existing research resources in the best possible ways. Cuts in research funds as a whole, or in specific types of funding, on the other hand, have led research and research fund applications to similarities among academic, governmental and private research units with regard to the type of work conducted.

Dr. Koskinen aptly pointed out that although there is and will be some overlapping academic, governmental and industrial research, each has its own differing responsibilities. At the same time it has to be kept in mind that in larger (national) programs all three units should be involved in cooperation. In addition, high priority should be given to information and communication flow as well as to international collaboration.

Better value for money through the joint research of various consortiums is one of the key ideas today. We have been fortunate to hear about three successful models functioning in different parts of the world: (1) Government (through national research laboratories) – industry – university practice in Japan presented by Dr. Sasaki. (2) A university – industry – government system created recently at Rutgers, the State University of New Jersey, by Dr. Solberg. (3) Research association – industry – government cooperation functioning in the United Kingdom and presented by Dr. Arthey.

Biotechnology is an old technology but progress in molecular biology has added new branches and new expectations to it. Biotechnology has been chosen as the priority area in the food sector in Japan. Extensive biotechnology research programs in the food and agricultural sector have also been carried out in Finland. Dr. Enari has illustrated for us some of the new opportunities created by biotechnology in the development of agricultural products used as feedstock in the food industry, industrial production processes, analytical methods and safety and quality assessment, and new food products.

Biotechnology is not, however, a panacea. There are several other techniques which have to be observed and carefully tested for potential exploitation. These include aseptic packaging, microwave and other high frequency and ohmic heating, membrane technology, combination technology, process control, and sensors.

Microorganisms are present in soil, water and air, so they are usually present in food and food raw materials. For instance, processing and proper storage are used to control undesirable growth of microbes in the product. Failures in proper conduction of present-day, established technologies and practices create problems and may result in illnesses and financial losses, as seen in various countries recently. New preservation technologies, on the other hand, may create new (micro)environments for microbial growth and new problems, as well as bringing about emerging pathogens. Hygiene, therefore, has always to be integrated into food technology, as observed by Dr. Nurmi.

Research in a small country like Finland can never fully cover all areas of interest. However, the country has to have a knowledge base wide and deep enough to enable exploitation of its own and foreign innovations. This is done partly by specialization in certain research areas where a center of excellence can be formed, partly by international cooperation. One of the extra benefits a small country can offer as a partner is a fresh approach and quick response both to problems and possibilities, as indicated by Dr. Olkku.

Food and its properties and quality, as well as attitudes and preferences of the consumer, greatly affect the buying behavior of the consumer and thus the success of the product and the company in the market. Other important factors to be considered are: progress of and results in science; possibilities and limitations resulting from food legislation; and policy of individual food companies as to food safety and quality. Dr. Paulus also showed us some results of the EEC Commission project on forecasting and assessment in science and technology.

Besides technical innovations in the food industry and innovations in related industrial fields and the subcontracting industry, important factors influencing the food industry in the years to come, as indicated by Dr. Spiess, may be social changes, changes in the ecological boundary conditions, new scientific knowledge of optimal forms of nutrition, and changes in raw materials.

We are now through with the lecture part of the minisymposium. I hope you have found it beneficial. As the Chairman of the Planning and Organization Committee, I want to express my deepest thanks to the speakers for very interesting and illustrative presentations, and to the other participants for their active discussion and comments. I hope the informal discussion and acquaintances already made will continue during the sauna and dinner. I consider this selected group of 40 participants from 11 countries to be a high level contact network which, let us hope, will become effectively utilized in the future for the mutual benefit of all parties involved.

Seppo Lindroth, PhD, Research and Development, Huhtamäki Oy Jalostaja,
PO Box 421, SF–20101 Turku (Finland)

Subject Index